D0074622

MAPPING WORKBOOK AND STUDY GUIDE FOR

WORLD REGIONAL GEOGRAPHY

Fourth Edition

Jennifer Rogalsky
State University of New York, College at Geneseo

Helen Ruth Aspaas
Virginia Commonwealth University

W. H. FREEMAN AND COMPANY
NEW YORK

Media and Supplements Editor: Deepa Chungi
Project Editor: Jodi Isman
Production Manager: Julia DeRosa
Map Designer: Will Fontanez
Marketing Director: Scott Guile

ISBN-13: 978-1-4292-0498-9
ISBN-10: 1-4292-0498-2

©2008 by W. H. Freeman and Company

All rights reserved.

Printed in the United States of America

First printing

W. H. Freeman and Company
41Madison Avenue
New York, NY 10010
Houndmills, Basingstoke R621 6XS, England
www.whfreeman.com

Contents

Preface v

Media Guide for Students vii

Chapter One. Geography: An Exploration of Connections 1

Chapter Two. North America 21

Chapter Three. Middle and South America 39

Chapter Four. Europe 61

Chapter Five. Russia and the Newly Independent States 81

Chapter Six. North Africa and Southwest Asia 101

Chapter Seven. Sub-Saharan Africa 119

Chapter Eight. South Asia 141

Chapter Nine. East Asia 159

Chapter Ten. Southeast Asia 177

Chapter Eleven. Oceania: Australia, New Zealand, and the Pacific 195

Appendix A: Answers to Sample Exam Questions

Appendix B: Blank World Maps

Preface

This workbook and study guide is designed to accompany the fourth edition of Lydia Pulsipher and Alex Pulsipher's *World Regional Geography*. The questions and exercises contain references to relevant tables and figures that appear in the fourth edition.

Each chapter of this workbook begins with **Learning Objectives** and **Key Terms**. These sections are intended for reference and for review of the main ideas of each chapter.

Your instructor may choose to use the **Review Questions** for class discussion or ask you to turn in written answers. You may decide to read through these questions before you read the textbook as signposts to main concepts in the text. These questions may also be used to review for an exam. Keep in mind that these questions cover only some of the material discussed in the textbook; they do not refer to every point made.

The **Critical Thinking Exercises** ask you to apply the concepts discussed in the textbook. These exercises may ask you to perform a task based on your daily routine, browse the Internet, or consider how your life might be different if you lived in another country. These exercises are designed to be stimulating and entertaining, while helping you to understand important geographic concepts.

Because places are so important in geography, you are given a list of **Important Places** for each region; most countries, capital cities, significant physical features, and other places discussed with some detail in the textbook are listed. You are asked to locate these places, and to note important facts about them.

It is likely that your instructor will ask you to turn in the **Mapping Exercises** as assignments. These exercises will help you to understand and explain geographic patterns. You are given blank maps to complete each exercise, and blank world maps for context. You can download additional copies of the maps from the textbook's Web site: www.whfreeman.com/pulsipher4e.

Finally, after working through the various components of this study guide, **Sample Exam Questions** are given for each chapter to test your knowledge of information and understanding of concepts from the textbook.

We hope you enjoy working through this student workbook. It should challenge and stimulate you to read the textbook thoroughly; it should also be fun, as you apply these situations to your own life and attempt to truly understand the complex geographic patterns and relationships of the world in which we live.

Enjoy!

JENNIFER ROGALSKY AND HELEN RUTH ASPAAS

Media Guide for Students

Web Site

www.whfreeman.com/pulsipher4e
Authors: Tim Oakes and Chris McMorran, University of Colorado, Boulder

For each of the 11 chapters in the textbook, the Web site serves as an online study guide, offering:

Map Learning Exercises: You can use these exercises to identify and learn about the countries, cities, and major geographic features of each region.

Thinking Geographically: This feature, presented for each chapter, encourages critical reflections on the linkages of trade, finance, tourism, and political movements that connect that region to the experience of residents of the United States. The activities in this section allow you to explore how a geographic perspective helps clarify our understanding of such issues as deforestation, human rights, and free trade. A number of links to a variety of Web sites that you can explore are included; these are matched with questions or brief activities that give you an opportunity to think about the ways your life is connected to the places and people you read about in the text. *Thinking Geographically* helps you focus on key concepts, such as scale, region, place, and interaction, by using these concepts to drive analysis of compelling issues. Icons in the text signal related *Thinking Geographically* exercises that you can work through on the Web site in order to more fully explore the textbook material and the concerns raised.

Working with Maps: You can develop your analytical abilities by working with two sets of map-related exercises:

- *Thematic Maps:* You can place various maps from the text side-by-side to compare and contrast data. Related questions accompany each option.
- *Animated Population Maps:* Animated maps show how regional populations have changed over time. Related questions ask how and why the changes may have occurred.

Blank Outline Maps: Printable maps of every text chapter region are provided for note-taking or exam review.

Online Quizzing: A self-quizzing feature enables you to review key text concepts and sharpen the ability to analyze geographic material for exam preparation. Your answers (correct or incorrect) prompt feedback, referring you to the specific section in the text where the question is covered.

Flashcards: A set of matching exercises that helps you learn vocabulary and definitions from the book's glossary is provided.

Audio Pronunciation Guide: A spoken guide for the pronunciation of place names, regional terms, and the names of historical figures is provided.

World Recipes and Cuisines (from *International Home Cooking*, the United Nations International School Cookbook): Class exercises and social events can be organized around the use of these recipes, through which you can gain an awareness and appreciation for international cuisine.

CHAPTER ONE
Geography: An Exploration of Connections

LEARNING OBJECTIVES

After reading the chapter and working through this study guide, you should:

- Know the key components of maps and the concepts associated with accurate map interpretation.
- Understand the guiding principles used in defining a region.
- Know why it is important to understand culture and its variety of cultural markers when examining a group's behavior and subsequent cultural change.
- Understand that gender roles differ across regions and cultures and that these roles significantly affect how a society works; identify reasons why gender roles differ among places and how they are changing.
- Know how different landforms evolve and how landforms are significant to climate patterns as well as human settlement patterns and agricultural practices.
- Know what it means to be part of the global economy. Understand that globalization is a dynamic process that has differential effects on rich and poor countries.
- Know how human well-being can be measured. Know the important differences among the ways of measuring well-being outlined in this textbook.
- Know how and why patterns of population growth and density differ from region to region. Understand some of the effects these patterns may have on gender and economic equality.
- Understand the concept of sustainable development as it relates to the environment, economics, culture, and human well-being.
- Understand the importance of political issues in geography, including geopolitics, nation-states, and international cooperation.

KEY TERMS

The following terms are in **bold** in the textbook. In the space next to the term (or on a separate sheet or flash cards), you can fill in the definitions for reference or quiz yourself for exam review. Definitions are found in the glossary of the textbook.

age distribution or age structure

air pressure

autonomy

average population density

birth rate

capital

carrying capacity

cash economy

climate

cold war era

colonialism

contiguous region

country

cultural diversity

cultural homogeneity

cultural identity

cultural marker

culture

culture group

death rate

delta

demographic transition

demography

deposition

development

"Development for whom?"

dialect

digital divide

domestication

earthquake

economy

erosion

ethnic group

European colonialism

exchange or service sector

external process (geophysical)

extraction

extractive resource

fair trade

floodplain

formal economy

formal institution

free trade

frontal precipitation

gender

gender structure

genocide

geopolitics

global economy

global scale

global warming

globalization

gross domestic product (GDP)

gross domestic product (GDP) per capita

growth rate

human geography

human well-being

import quota

industrial production

Industrial Revolution

informal economy

informal institution

institution

internal process (geophysical)

International Monetary Fund (IMF)

interregional linkage

landform

latitude

lingua franca

living wage

local scale

longitude

material culture

monsoon

multiculturalism

multinational corporation

nation

national identity

nation-state

Neolithic Revolution

nongovernmental organization (NGO)

nonmaterial resource

norm

Organization for Economic Cooperation and Development (OECD)

orographic rainfall

Pangaea hypothesis

physical geography

plate tectonics

pluralistic state

political ecologist

population pyramid

purchasing power parity (PPP)

race

racism

rain shadow

rate of natural increase (growth rate)

region

regional geography

regional trade bloc

religion

remittance

resource

Ring of Fire

scale

secularism

sovereignty

spatial analysis

structural adjustment policies (SAPs)

subduction

subregion

subsistence economy

sustainable agriculture

sustainable development

tariff

technology

total fertility rate (TFR)

United Nations (UN)

volcano

weather

weathering

World Bank

world region

World Trade Organization (WTO)

REVIEW QUESTIONS

The following review questions are related directly to the textbook material. These questions can be used to help you prepare for an exam or you may want to read through the questions before you begin reading the textbook, quizzing yourself after you complete each section.

Introduction

1. List at least three challenges that geographers encounter as they designate the regions of the world.
2. Identify at least three potential impacts of the arrival of a manufacturing or assembly plant in a highly traditional culture. What are the risks and opportunities associated with this connection to globalization?

Cultural and Social Geographic Issues

3. How is language related to culture? What is happening to the diversity of languages and what effect will this have on cultures in the future?
4. Why has the English language evolved as the global lingua franca?
5. Examine the photos in Figure 1.18 (the home and its contents in California and in Mongolia) and use them to explain the material culture, technology, and cultural markers for each of the two cultures.
6. Although activities assigned to men and women differ among cultures and eras, what are some of the consistencies? Attempt to explain the situations in which education levels are higher for females than for males (Table 1.1).
7. Although race is not seen as biologically significant, why does it have political and social importance?

Physical Geography: Perspectives on the Earth

8. How can you account for Africa's general plateau-like surface?
9. Explain the process that creates monsoons. Include a discussion of temperature and air pressure, as well as differential heating and cooling of water.

10. Explain the tie between domestication of plants and trade.

Economic Issues in Geography

11. How can workers who live in one country earn relatively high wages when compared to workers in another country, yet still live in poverty?
12. How did European mechanization affect colonies' abilities to become producers of finished products?
13. How are the effects of the growth of multinational corporations similar to the effects of colonization?

Measures of Development

14. Identify the pros and cons of using each of the following as a development measure: Gross Domestic Product per capita, United Nations Human Development Index, United Nations Gender Development Index, and United Nations Gender Empowerment Measure.
15. Suggest a situation that can lead to a country having a low GDP but a high HDI and GDI.

Population Patterns

16. Relate the J curve to the growth of human populations over the last 2000 years.
17. If you observe a population pyramid for a country that shows fewer girls and women than boys and men, what are some of the conclusions you can draw about that country's gender issues?

Humans and the Environment

18. Why is soil preservation such a vital aspect of sustainable development?
19. Why is the relatively rich minority population causing such problems for the environment? How is this pattern changing as the developing countries industrialize, modernize, and urbanize? What are the probable impacts on the environment?

Political Issues in Geography

20. Identify those arguments that support the concept of state sovereignty. Likewise, identify those arguments that criticize the concept of state sovereignty.

CRITICAL THINKING EXERCISES

The following are "what if"-type questions that illustrate concepts you are learning from the textbook. These questions ask you to apply the ideas and principles you learned from the textbook to new situations.

1. Geography and its relations to other disciplines

Geography can be defined as the study of our planet's surface and the processes that shape it. However, geographers usually specialize in one or more fields of study.

- Consider your academic field of interest. How does it fit into geography?
- In what ways does your field of interest fit into one of the geography subdisciplines listed in the "What is Geography" section of your textbook?

2. The information revolution

To be a revolution of the same magnitude as the industrial or agricultural revolution, the information revolution would have to transform the ways in which people relate to their environments and to one another. Just as in the agricultural and industrial revolutions, not all people have been affected by the information revolution. Therefore, we see people today who are on the negative side of the digital divide.

- Discuss whether you believe that the information revolution is on a par with the two previous major technological revolutions. Defend your response.
- Identify at least five methods that could be used to narrow the digital divide.

3. Values and ways of knowing

All cultures establish, preserve, and pass on knowledge, which is grounded in a set of values. These values differ from culture to culture (and sometimes within cultures); thus, a type of behavior might be admired by some but reviled by others.

- Select a non-Western cultural group that is represented on your campus. Ask a member of that group to identify some of his or her culture's values.
- Identify similarities between your culture's values and those of the non-Western culture. Identify differences.
- Suggest a generalization about why there are similarities but distinguishing differences between Western cultures and non-Western cultures.
- Are there overarching values or standards among cultures?
- How can we be sensitive to differences among places and within cultures regarding larger issues of human rights (oppression, genocide, torture, etc.)?

4. Pacific Ring of Fire

The Pacific Ring of Fire is an especially active area of volcanoes and earthquakes. In recent years, it has received worldwide attention, as hundreds of thousands of people have died because of tectonic activity (and the resulting tsunamis).

- Examine Figure 1.21 (Ring of Fire), which includes locations of earthquakes, volcanoes, and plate boundaries.

- Visit earthquake.usgs.gov/regional/world/historical_country.php for a list of recent earthquakes. Click on the hyperlinks to view maps and read about the events.
- volcano.und.nodak.edu/vwdocs/volc_images/southeast_asia/southeast_asia.html allows you to click on the volcano icons to see images of and read about recent volcano eruptions.
- Using the Internet, find the Banda Aceh region and the epicenter of the 2004 Indian Ocean tsunami: www.news.ucdavis.edu/special_reports/tsunami or www.cbc.ca/news/background/asia_earthquake/gfx/map_epicentre.jpg
- Reflect on what you've seen on these Web sites. What would life be like if you lived in the Ring of Fire? What methods could you use to adapt and cope with these potentially life-threatening events?

5. Formal and informal economies

The formal economy includes all the activities that are recorded as part of a country's official production. The informal economy, however, includes goods and services that are produced outside formal markets, often for no cash or for payment that is not reported to the government.

- Consider the type of "informal economy" employment you have had in the last five years. Why was this work *not* part of the formal economy?
- What are/were the positive effects of this on your life and the lives of others?
- What are/were the negative effects of this on your life and the lives of others?
- Because of educational, racial, or gender restrictions, imagine that you were able to only be part of the informal economy. How would this affect the quality of your life in the past, present, and the future?
- What would your life be like if you could not do/have done any work in the informal economy?

MAPPING EXERCISES

The following are three mapping exercises to improve your knowledge of the location of places, underscore why they are important, and clarify how they relate to each other. Some questions will ask you to locate places, compare maps, or fill in data; others will test your understanding of *why* you were asked to map the features that you did. Use the blank outline maps at the end of the chapter to complete these exercises. Additional blank outline maps can be found on the textbook's Web site: www.whfreeman.com/pulsipher4e.

1. Is there a digital divide between North America and Central America?

The digital divide refers to the division between people who use computers and have access to the Internet and those who do not.

- Use the blank maps for North America and Central America to complete this mapping exercise.

- Visit the CIA World Factbook at www.cia.gov/cia/publications/factbook.
- Obtain information for the GDP per capita and the number of Internet users for each of the following countries: Canada, the United States, Mexico, Belize, Guatemala, Honduras, Nicaragua, El Salvador, Costa Rica, and Panama.
- Choose two different symbols to map the GDP per capita and the number of Internet users. Shade each country with a shade of gray (darkest shade of gray representing the highest GDP per capita). Use graduated circles to portray the number of Internet users.

Questions

a. What is the relationship between the GDP per capita for each country and its number of Internet users? Explain the relationship.
b. Based on your maps, can you say that the digital divide exists? If it does, is there a clear division between those countries that are rich and have access to the Internet and those countries that are poor and have minimum access to the Internet? If not, try to explain why this is the case.
c. For poor countries, how would stepping across the digital divide and gaining more Internet access help improve living conditions for the poor?

2. Arable land, agriculture, and soil degradation

Soil erosion appears to be an environmental concern in many countries, but it may be more serious in those countries where the economy has a high dependency on agriculture. All the countries in South Asia are facing soil degradation issues.

- Visit the CIA World Factbook at www.cia.gov/cia/publications/factbook.
- Collect information on percent of arable land, percent of population who earn their living from agriculture, and the GDP per capita for the countries of South Asia: India, Pakistan, Bangladesh, Afghanistan, Nepal, Bhutan, and Sri Lanka.
- Select appropriate symbols to map these three sets of information on the blank map of South Asia. You may want to use shading for one variable, cross-hatching for another variable, and a symbol of your choice for the third variable.

Questions

a. From your map, identify at least three challenges that farmers face if they want to slow the rate of soil degradation occurring on their farms.
b. Identify the country that may have the most serious challenges in arresting soil degradation and explain why you selected that particular country.

3. Climate classifications

Climate, wind, and weather are largely the result of complex patterns of air temperature and pressure, ocean temperature and currents, and the location of certain landforms.

Questions

a. Consider the climate of where you currently live. List the general climate characteristics including temperature, precipitation, and seasonality characteristics.

b. Next look at Figure 1.22 (precipitation), and Figure 1.25 (climate regions). Write down the description or classification from each figure that corresponds to your area.
c. If the description or classification matches how you described your area, attempt to explain your climate characteristics. If the description or classification does not match yours well, explain why it does not match. Consider the scale of the maps in Figures 1.22 and 1.25, as well as any mountain ranges or bodies of water that may affect climate.

SAMPLE EXAM QUESTIONS

The following are sample questions to help you review for an exam. Answers are found in the back of this workbook.

1. The term *secularism* describes what type of society?
   A. One in which all forms of exchange occur through barter
   B. One in which power is shared by multiple ethnic groups
   C. One in which the way of life is not directly informed by religious values
   D. One in which external powers exert control over the political system

2. The term *lingua franca* refers to which of the following?
   A. A language that was once extinct but has been revived
   B. Language formed of various parts of different languages
   C. A language in the Indo-European family of languages
   D. The dominant, if not universal, language used in trade

3. Researchers in which of the following fields study the processes by which the physical landscape is shaped?
   A. Climatology
   B. Cartography
   C. Geomorphology
   D. Biogeography

4. Which of the following events is NOT directly associated with plate tectonics?
   A. Earthquakes
   B. Fluvial deposition
   C. Volcanic eruptions
   D. Mountain building

5. What term describes rainfall that is caused by the interaction of large air masses of different temperatures?
   A. Orographic precipitation
   B. Shadow precipitation

C. Convergence precipitation
D. Frontal precipitation

6. Which of the following conveys the meaning of the term *sustainable development*?
A. The ability to generate ways of increasing economic growth that can be maintained in the future
B. The establishment of free markets through which each person has the ability to improve his or her life
C. The endeavor to improve economic practices that provide only sporadic growth and wealth to individuals
D. The effort to improve present standards of living without jeopardizing those of future generations

7. Which of the following accords with the theory of *global warming*?
A. Deforestation reduces the release of carbon dioxide causing more atmospheric combustion.
B. The burning of fossil fuels decreases levels of carbon dioxide preventing natural ionic cooling.
C. The sun is burning hotter and has increased rates of evaporation, and thus has a reduced rate of global daily cooling.
D. Levels of carbon dioxide in the atmosphere are increasing and trapping the sun's heat.

8. Compared to a society in which most people work in agriculture, which of the following is true about the resource consumption of people in a society composed of industrial and service workers?
A. They draw resources from a much wider area.
B. They use resources less rapidly.
C. They consume fewer resources overall.
D. They generally pay higher prices.

9. Which of the following relationships is accounted for in the rate of natural increase?
A. Birth rates and death rates in a given population
B. Immigrants and emigrants to and from a given population
C. Death of infants per 1000 born
D. Newborn infants and migrants added to a population in a given year

10. Which of the following activities is most likely to occur in the *informal* economy?
A. Tourism
B. Housework
C. Logging
D. Transportation services

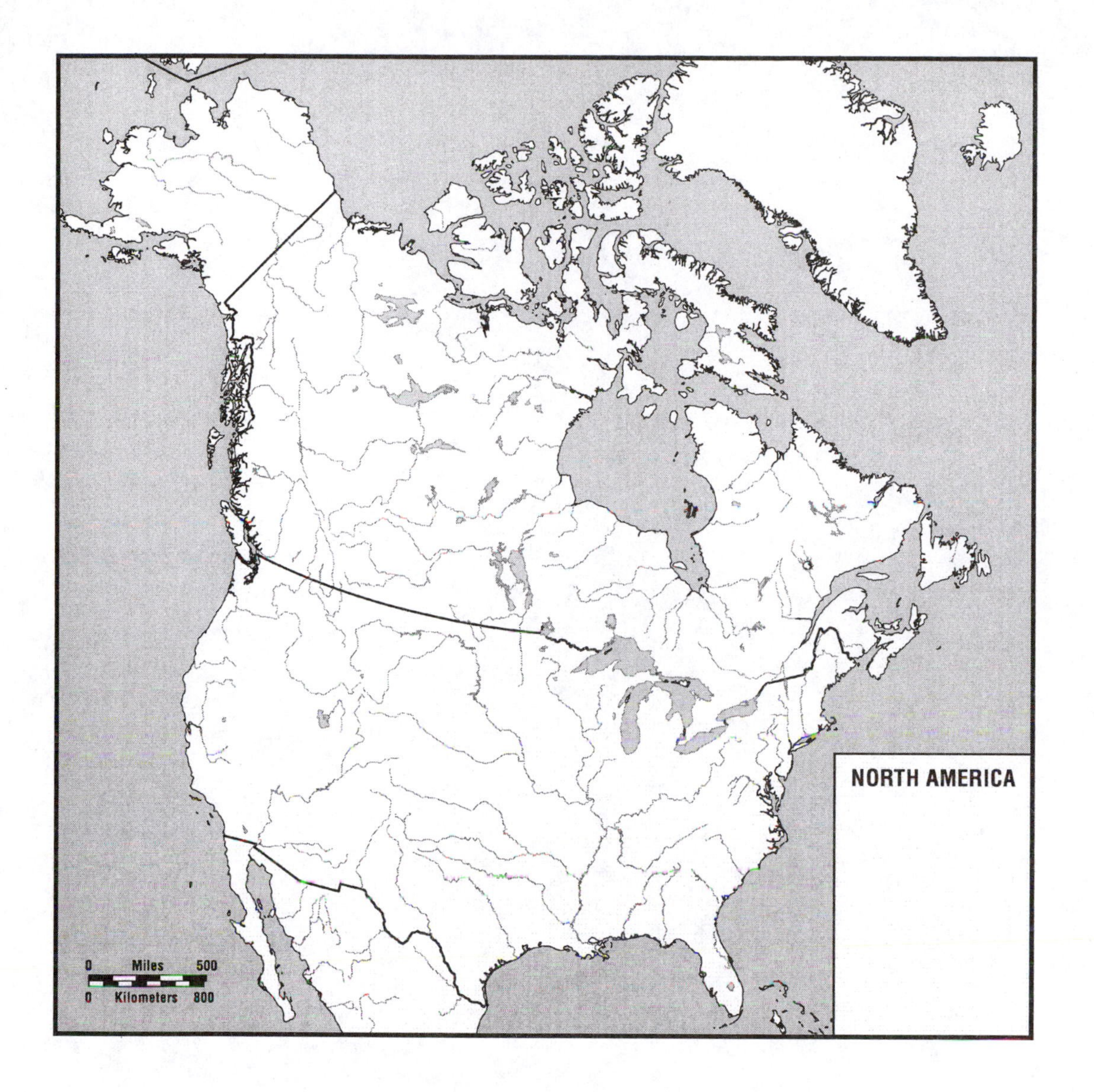
NORTH AMERICA
0 Miles 500
0 Kilometers 800

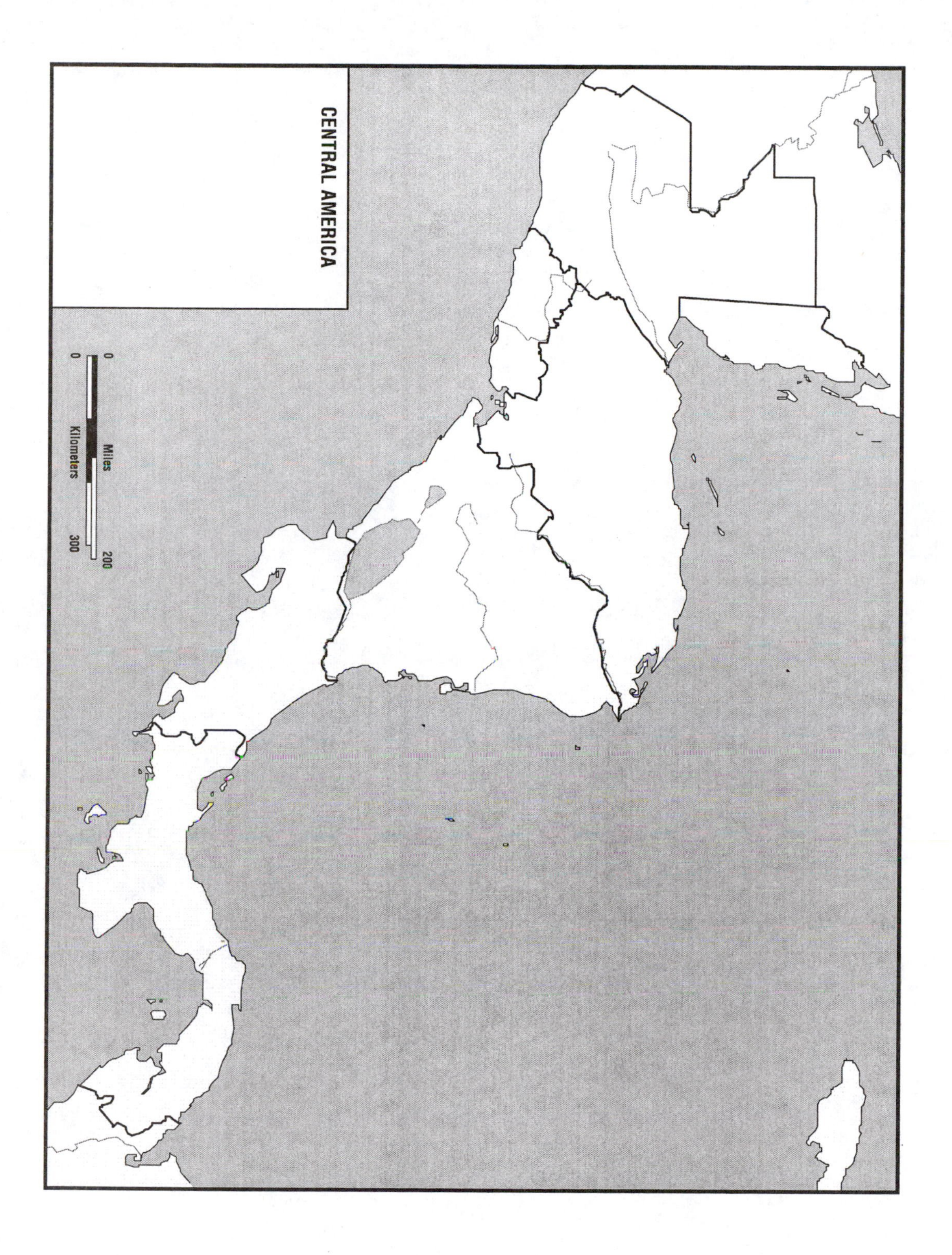
CENTRAL AMERICA
Miles
0
200
Kilometers
0
300

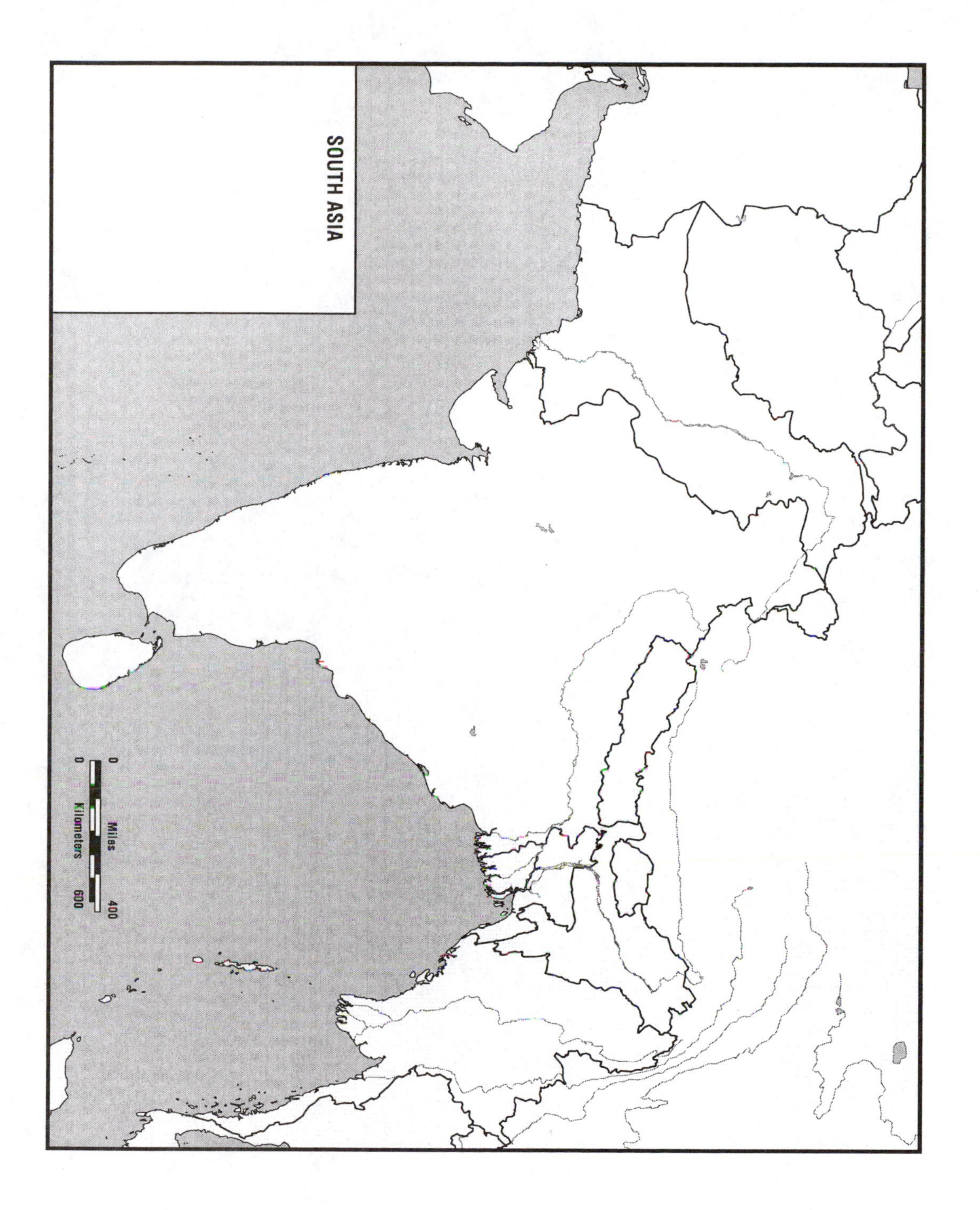
SOUTH ASIA
0
Miles
400
0
Kilometers
600

CHAPTER TWO
North America

LEARNING OBJECTIVES

After reading the chapter and working through this study guide, you should:

- Know that North America is comprised of two large countries and understand how diverse and complex landforms are often associated with a variety of climatic patterns that in turn affect settlement patterns and agricultural practices.
- Know the historical and contemporary processes that transform population densities and the ethnic diversity of North America.
- Know that despite similarities in wealth, governance, and standards of living, Canada and the United States are asymmetric in population size, total economic output, and roles in international affairs.
- Know that the region is experiencing economic transition from agriculture and industry to postindustrial service and technology economies, many of which are associated with a highly mobile and well-educated workforce.
- Know that although women are attaining career opportunities, they frequently are faced with the challenges of coping with family responsibilities, rapidly evolving household structures, and issues arising from the income disparity between themselves and their male counterparts.
- Know that social, political, and economic interactions between this region and the rest of the world are fraught with positive and negative repercussions, especially in the context of world trade, immigration, and national security.
- Know that the economic vitality and standard of living for this region have come at high cost to the environment.

KEY TERMS

The following terms are in **bold** in the textbook. In the space next to the term (or on a separate sheet or flash cards), you can fill in the definitions for reference or quiz yourself for exam review. Definitions are found in the glossary of the textbook.

acid rain

agribusiness

aquifer

baby boomer

boreal forest

brownfield

Canadian Shield

chain migration

clear-cutting

dense node

digital divide

economic core

ethnicity

genetic engineering

Geneva Conventions

gentrification

government subsidy

Great Basin

hazardous waste

Hispanic

hub-and-spoke network

information technology (IT)

Internet

Interstate Highway System

knowledge economy

Latino

loess

long-lot system

maquiladora

megalopolis

metropolitan area

New Urbanism

North American Free Trade Agreement (NAFTA)

nuclear family

Ogallala aquifer

outsourcing

Pacific Rim (Basin)

permeable national border

Québecois

Rust Belt

service sector

smog

taiga

thermal inversion

trade deficit

tundra

urban sprawl

REVIEW QUESTIONS

The following review questions are related directly to the textbook material. These questions can be used to help you prepare for an exam or you may want to read through the questions before you begin reading the textbook, quizzing yourself after you complete each section.

The Geographic Setting

1. What are the geologic processes responsible for the physical landscape of the three major landforms: the Rocky Mountains, the Appalachian Mountains, and the Central Lowland?
2. Winds, mountain ranges, and water bodies account for prevailing weather conditions and climatic patterns in much of North America. What are the general impacts of each of the major water bodies on regional weather conditions?
3. What was the impact of European colonization and expansion on Native American populations in terms of land access and indigenous property holdings?
4. Name five key events that have affected the ethnic and geographic redistribution of population in this region. What are the long-term impacts of the population shift from northeastern United States to the Southeast and the West?

Current Geographic Issues

5. Name at least two divisive issues associated with the increasing focus on national security within the United States. Explain the reasons for these debates.
6. Name two criteria that suggest that Canada's dependency on the United States is much greater than the dependency of the United States on Canada. Explain how Canada is able to maintain its autonomy, despite having a powerful neighbor along its southern border.
7. What major changes have taken place in the relationship between agriculture and national economies, employment, and markets during the twentieth century and continuing into the twenty-first century? What are the impacts of these changes on the consumer, both domestically and internationally?
8. Compare agriculture, manufacturing, and services to each other in terms of share of the economy and percent of labor force involvement. Relate your findings to existing human resources in the region.
9. How has this region used global connections to further its economic interests?
10. Despite increased opportunities for women in North America, many aspects of men's and women's lives are not at parity. What are some of these differences? How can we account for differences between Canada and the United States?
11. What are three problematic processes associated with urbanization? Explain programs and new policies enacted to alleviate these problems.
12. How do immigrants take from and contribute back to their communities and economies?
13. What are the challenges that individual races and ethnic groups face as this region becomes more homogeneous in the context of culture and urbanization?

14. What factors led to the evolution of the nuclear family? What factors appear to be influencing its demise in the twenty-first century?
15. What are the challenges that individual "baby boomers" and society as a whole will face as the "baby boomers" move into retirement age?
16. Figure 2.37 (changing distribution of the elderly in the United States) predicts the rise in the elderly population in states like Florida and Arizona, which can be attributed to retirement communities. How can we account for the estimated increase in the elderly population for the states that comprise the Breadbasket?
17. What factors have promoted a high incidence of mobility among this region's population? What are some long-term impacts on the social fabric of the region?
18. How does the choice of location for hazardous waste disposal affect certain economic and ethnic groups in the United States?
19. Name at least three factors that are contributing to the high rate of flora and fauna extinction in this region. Suggest some solutions that will help slow this process.
20. Using the indicators of human well-being, make an argument that Canada's population may surpass that of the United States in terms of overall well-being.

CRITICAL THINKING EXERCISES

The following are "what if"-type questions that illustrate concepts you are learning from the textbook. These questions ask you to apply the ideas and principles you learned from the textbook to new situations.

1. Hispanic influences on American culture

Hispanics comprise the principal source of immigration into the United States. Figure 2.2 (foreign-born residents) and Figure 2.27 (changing national origins of legal U.S. immigrants) give visual evidence of the immigration of Hispanics into North America.

- Make at least three observations regarding the impact of an increased Hispanic population presence on American culture and lifestyles today. For example, some sources suggest that salsa has replaced ketchup as the condiment of choice on the average American table.
- How have some of your personal activities and preferences changed as a result of this influence?
- Suggest three impacts that might result from continued immigration of Hispanics into the United States during the next decade or two. If your impacts imply a problematic situation, then suggest means of alleviating them.
- What are some of the most strident arguments against immigration that you have heard around campus or in your neighborhood? Likewise, what are some of the most supportive arguments favoring immigration that you have heard recently?
- After examining these two sides of the debate, reach some form of consensus that balances both perspectives.

2. Americans and their homogeneous landscape

The authors suggest that Americans have created a homogenous landscape as an attempt to help the highly mobile population always feel "at home."

- Think about your city, town, or community and list at least five structures that are part of the universal American landscape.
- Identify the location of these structures in relation to the city center and evaluate the role of the automobile in helping to create this homogeneous landscape.
- Describe your use of these structures and your rationale for doing so.

3. Changing family structures in North America

The family structure in North America experienced considerable change during the last half of the twentieth century.

- Examine your own family's structure as it exists today and compare it to that of your parents when they were growing up, and also to that of your grandparents when they were children.
- Using these case studies, determine if your family structure for the past three generations matches the changes over time in American family structure as discussed in the text.
- Examine Figure 2.33 (U.S. households by type) and relate your family structure to this figure.

4. Aging: Different perspectives from different generations

Different generations may respond to issues associated with aging in different ways.

- Compare your perceptions of aging to that of your parents.
- If possible, ask your parents what their expectations were of "old age" when they were in college or just beginning their careers. Ask your parents how their expectations of "old age" and retirement have changed as they have aged.
- What are your own expectations of "old age" now?
- Compare and contrast your findings to those in the text.
- While population growth increases environmental impacts, slower population growth increases the population that is old, leaving fewer, young working people to shoulder the demands of a growing economy. In light of this, should population growth be curtailed?
    - Place the justification for slower growth against the need for young workers in North America and provide some workable solutions.

5. Mobility and suburbia in North America

One principal characteristic of the North American population is the frequency with which it moves about. Many of these moves are to the suburbs, despite growing concerns about urban sprawl and loss of valuable agricultural land.

- Reflect on your own experiences with moving both as you were growing up and as you entered college. Describe the rationale for each of those moves.
- How does your own history of mobility compare to that discussed in the chapter?
- What perceptions of both the central city and the suburbs do people in the United

States and Canada have that encourage the movement to suburbia? What local, state, and federal policies *encourage* the move to suburbia?

- What benefits are accrued to society by enhancing urban areas and retarding the rush to the suburbs? What ideas can you generate that could entice suburban residents back to the cities?

IMPORTANT PLACES

The following places are featured in the chapter. Make sure you can locate all of them on a map. Blank outline maps can be found on the textbook's Web site: www.whfreeman.com/pulsipher4e. Also, to prepare for quizzes and exams, write a few important facts about each place in the space provided.

Physical Features

1. Appalachian Mountains
2. Atlantic Ocean
3. Bering land bridge
4. Canadian Shield
5. Catskill-Delaware watershed
6. Colorado River
7. Columbia River
8. Grand Banks
9. Great Basin
10. Great Lakes
11. Great Plains
12. Gulf of Mexico
13. Hudson Bay
14. James Bay
15. Lake Michigan

16. Mississippi River Valley and Delta

17. Missouri River

18. Ogallala Aquifer

19. Ohio River

20. Pacific Ocean

21. Peace River

22. Rocky Mountains

23. Snake River

24. St. Lawrence River

25. Utah Valley

Regions/Countries/States/Provinces

26. Alabama

27. Alaska

28. Alberta

29. Appalachia

30. Arizona

31. Arkansas

32. Atlantic Provinces

33. British Columbia

34. California

35. Colorado

36. Connecticut

37. Continental Interior

38. Delaware

39. Florida

40. Georgia

41. Great Plains Breadbasket

42. Idaho

43. Illinois

44. Indiana

45. Iowa

46. Kansas

47. Kentucky

48. Louisiana

49. Maine

50. Manitoba

51. Maritime Provinces

52. Maryland

53. Massachusetts

54. Michigan

55. Midwest (Middle West)

56. Minnesota

57. Mississippi

58. Missouri

59. Montana

60. Nebraska

61. Nevada

62. New Brunswick

63. New England

64. Newfoundland

65. New Hampshire

66. New Jersey

67. New Mexico

68. New York

69. North Carolina

70. North Dakota

71. Northeast

72. Northwest Territories

73. Nova Scotia

74. Nunavut

75. Ohio

76. Oklahoma

77. Old Economic Core

78. Ontario

79. Oregon

80. Pacific Northwest

81. Pennsylvania

82. Prairie Provinces

83. Prince Edward Island

84. Quebec

85. Rhode Island

86. Saskatchewan

87. South Carolina

88. South Dakota

89. Southeast (The American South)

90. Southwest

91. Tennessee

92. Texas

93. Utah

94. Vermont

95. Virginia

96. Yukon Territories

97. Washington

98. The (American) West

100. West Coast

101. West Virginia

102. Wisconsin

103. Wyoming

Cities/Urban Areas

104. Atlanta, GA

105. Austin, TX

106. Baltimore, MD

107. Bentonville, AR

108. Birmingham, AL

109. Boston, MA

110. Calgary, Alberta

111. Chicago, IL

112. Cincinnati, OH

113. Cleveland, OH

114. Dallas, TX

115. Denver, CO

116. Detroit, MI

117. El Paso, TX

118. Eugene, OR

119. Georgetown, KY

120. Hackensack, NJ

121. Hartford, CT

122. Indianapolis, IN

123. Jamestown, VA

124. Kansas City, MO

125. Laramie, WY

126. Lewiston, ME

127. Laredo, TX

128. Los Angeles, CA

129. Lowell, MA

130. Megalopolis

131. Miami, FL

132. Minneapolis, MN

133. Montreal, Quebec

134. Nashville, TN

135. New Orleans, LA

136. New York, NY

137. Nogales, AZ

138. North Lawndale, IL

139. Philadelphia, PA

140. Phoenix, AZ

141. Pittsburgh, PA

142. Pittsford, NY

143. Portland, OR

144. Providence, RI

145. Québec City, Québec

146. Raleigh-Durham, NC

147. Rochester, NY

148. Salt Lake City, UT

149. San Diego, CA

150. San Francisco, CA

151. Seattle, WA

152. St. Louis, MO

153. Storm Lake, IA

154. Terrebone Parish, LA

155. Toronto, Ontario

156. Valdez, AL

157. Vancouver, British Columbia

158. Washington, D.C.

MAPPING EXERCISES

The following are three mapping exercises to improve your knowledge of the location of places, underscore why they are important, and clarify how they relate to each other. Some questions will ask you to locate places, compare maps, or fill in data; others will test your understanding of *why* you were asked to map the features that you did. Use the blank outline maps at the end of the chapter to complete these exercises. Additional blank outline maps can be found on the textbook's Web site: www.whfreeman.com/pulsipher4e.

1. Regional climate, population density, and agriculture practices

Agriculture in North America is highly productive, but usually in areas with favorable climates and abundant natural resources. Refer to the climate map (Figure 2.6), the population distribution map (Figure 2.12), and the agriculture map (Figure 2.18) to answer the following questions.

Questions

a. What are two possible human and two physical geography explanations for the location of the *mixed farming* activities?
b. What are two possible human and physical geography explanations for the location of the *range livestock* activities?
c. What are two possible human and physical geography explanations for the location of *corn belt, cash grain,* and *livestock* activities?

2. Mobility and aging in developed societies

For a multitude of reasons, the populations of this region are highly mobile. In some cases, mobility can increase as one ages, especially for those eager to enjoy the amenities of warm climates, say in Arizona or Florida. Use the population by region map (Figure 2.13) and the changing distribution of elderly map (Figure 2.37) to answer the following questions.

Questions

a. Which regions of the United States have experienced the largest increase in population between the years of 1900 and 2000? Provide at least three reasons to explain this noteworthy redistribution of the U.S. population.
b. It was noted previously that Arizona and Florida attract many senior citizens who move to these states to enjoy retirement. How, then, do we explain the growing population of elderly in the Great Plains Breadbasket, the Rust Belt, and New England?

3. Environmental issues in North America

North America's environmental issues vary across the region. Some issues are highly localized while others affect large populations. Refer to the air and water pollution map (Figure 2.38), the population density map (Figure 2.12), the agriculture map (Figure 2.18), and the human impact map (Figure 2.39) to answer the following questions.

Questions

a. Which environmental issues are likely to be associated with high population densities? Why is this the case?
b. Which environmental issues are not likely to be associated with high population densities? Why is this the case?

SAMPLE EXAM QUESTIONS

The following are sample questions to help you review for an exam. Answers are found in the back of this workbook.

1. Which of the following is true regarding the situation of women in North America?
   A. Women now comprise about half the labor force.
   B. On average, women earn 45 percent of what men earn for out-of-home work.
   C. The percentage of women in national legislatures is the world's highest.
   D. Women own a majority of businesses.

2. According to the 2000 census, what is the largest minority group in the United States?
   A. African-Americans
   B. Hispanics
   C. Asian-Americans
   D. Native Americans

3. Which of the following climate types is NOT found in North America?
   A. Arctic
   B. Desert
   C. Temperate
   D. Tropical

4. According to the text, which of the following is mentioned as a characteristic of most new immigrants to the United States?
   A. They commit crimes out of desperation.
   B. They tend to pay taxes.
   C. They use more public services than they pay for (through taxes).
   D. They are usually unemployed several months to years after arrival.

5. Which of the following statements describes agriculture in North America?
   A. Agriculture is a growing source of employment.
   B. The agriculture system is an important producer of food for domestic and foreign consumers.
   C. Agriculture has been strictly private, receiving little government aid.
   D. Most large, corporate farms have gone bankrupt; family-owned farms now produce 75 percent of the agricultural needs of North America.

6. Which of the following is true regarding the high-tech industry today?
   A. It generally depends on a pool of low-skilled labor.
   B. It has led to the increased economic development of previously rural areas.
   C. It is considered part of the manufacturing/industrial economic sector.
   D. Businesses are often located near major universities or research institutions.

7. The followers of what version of Christianity predominate in the so-called "Bible Belt" of the United States?
   A. Lutheran
   B. Mormon
   C. Baptist
   D. Catholic

8. Which of the following is true regarding the U.S. and Canadian economies?
   A. The United States is more dependent on Canada than vice versa.
   B. Canada is more dependent on the United States than vice versa.
   C. Both trade more with Japan than each other.
   D. Both trade more with Europe than each other.

9. Which of the following is true regarding plant and animal species in North America?
   A. Wetlands have been filled in to provide more suitable habitat for endangered bird species.
   B. Exotic plant and animal species are helping native species recover by stabilizing native habitats.
   C. The United States has the largest number of plant species threatened with extinction.

D. Although past development in early North American history resulted in many threatened and endangered species, almost all of these species no longer need special protection.

10. The passage of the North American Free Trade Act (NAFTA) has had which of the following effects?
    A. Considerable decline in trade between the United States and Canada.
    B. Reduction and removal of tariffs between Mexico, the United States, and Canada.
    C. Rising corruption in Canada due to United States and Mexican firms who have relocated there.
    D. Sharp decline in the flow of Mexican migrants into North America.

CHAPTER THREE
Middle and South America

LEARNING OBJECTIVES

After reading the chapter and working through this study guide, you should:

- Know basic landform and climate patterns, including the tectonic forces in the region and how climate varies across the large land and island landmass.
- Understand the influence of colonization on the landscape, settlement patterns, economy, and populations of countries in the region.
- Understand the general patterns of population distribution. Know why some countries are still growing rapidly while others are growing quite slowly.
- Know the causes and effects of structural adjustment programs, the overwhelming debt that many countries are facing, and free trade agreements in this region.
- Understand why democracy is still so fragile in this region and what can be done to make it succeed.
- Understand the importance of family life, especially the extended family. Understand the importance and consequences of *machismo* and *marianismo*.
- Know the environmental issues that are threatening ecosystems in this region. Understand how economic development and the environment are linked.

KEY TERMS

The following terms are in **bold** in the textbook. In the space next to the term (or on a separate sheet or flash cards), you can fill in the definitions for reference or quiz yourself for exam review. Definitions are found in the glossary of the textbook.

acculturation

altiplano

assimilation

Aztecs

barriadas

barrios

brain drain

colonias

contested space

conurbation

Creoles

Creole cultures

early extractive phase

economic restructuring

ecotourism

El Niño

evangelical Protestantism

Export Processing Zones (EPZs)

extended family

external debts

favelas

foreign direct investment (FDI)

forward capital

hacienda

import substitution industrialization (ISI)

Incas

income disparity

indigenous

isthmus

ladino

land reform

liberation theology

machismo

maquiladoras

marianismo

mercantilism

Mercosur

mestizos

Middle America

mother country

neocolonialism

North American Free Trade Agreement (NAFTA)

plantation

populist movements

primate city

pyroclastic flow

shifting cultivation

silt

South America

subduction zone

subsidence

temperature-altitude zones

tierra caliente

tierra fria

tierra helada

tierra templada

trade winds

urban growth poles

REVIEW QUESTIONS

The following review questions are related directly to the textbook material. These questions can be used to help you prepare for an exam or you may want to read through the questions before you begin reading the textbook, quizzing yourself after you complete each section.

The Geographic Setting

1. How do plate tectonics affect physical geography in this region?
2. Explain why this region has such climatic diversity.
3. What are some of the known effects of El Niño and hurricanes on this region?
4. What were the strengths of the Aztecs and Incas during the conquest? How did the Spanish defeat them, given the relatively small number of Spanish soldiers?
5. What is the basic relationship between a mother country and its colonies in Middle and South America? In what ways were the populations of the mother countries and the colonies affected by colonization?
6. What are the positive and negative consequences of population growth? Why do some countries have such high rates of natural increase, while others are quite low? What does rapid growth mean for economies and human well-being?
7. What are some of the reasons for rural to urban migration? What are some of the outcomes in both the origin (rural villages) and the destination (cities)?

Current Geographic Issues

8. Give a brief description of the three stages of economic development as presented in the textbook. Include information on the type of industry common in each stage, the impact of government intervention, and the effects on the economy and population.

9. Explain the differences between haciendas, plantations, and ranches. Where are each of them commonly located? What is the typical worker/owner relationship?
10. If a country has an import substitution policy, what kinds of things do you think it may try to manufacture? What are the advantages and disadvantages?
11. What is meant by the term "debt crisis," and why has it been determined that structural adjustment programs (SAPs) will help? What are the impacts of the SAPs and current debt at the personal, national, and regional levels?
12. What factors have led to widespread participation in the informal economy? What are the positive and negative outcomes on individuals and economies?
13. What are the positive and negative impacts of free trade agreements at the personal, provincial, and national scale?
14. Many factors can compromise the success of establishing and maintaining a democracy. What are some obstacles to stability and democracy in Middle and South America?
15. What are the roles, reasons for, and effects of *machismo* and *marianismo*? What transitions are occurring in both of these roles in the twenty-first century?
16. What are some differences between extended families and nuclear families? How does an extended family affect the arrangement of living space, relationships between parents and children, and spending of family income?
17. Explain how liberation theology and Evangelical Protestantism address the matter of wealth and poverty in quite different ways.
18. Many are concerned about the fast pace of rain forest loss. What are some human activities that result in deforestation? How is deforestation linked to the disparity of wealth?
19. Why is ecotourism an appealing development idea for some countries? What may be some of the drawbacks?
20. What circumstances could explain why gross domestic product (GDP) per capita might be relatively low for a country, yet its Human Development Index (HDI) and Gender Empowerment Measure (GEM) ranks might be high? What factors form the basis for saying that the quality of life is relatively high in the Caribbean?

CRITICAL THINKING EXERCISES

The following are "what if"-type questions that illustrate concepts you are learning from the textbook. These questions ask you to apply the ideas and principles you learned from the textbook to new situations.

1. Migration push and pull factors

Migration is one of the most important social forces in the world today. Many economic and social factors initiate and sustain migration, and likewise it has many social and economic effects.

- Consider where you live or attend college now. What are the reasons you left the last place you lived (push factors)?
- Have you ever had to move because of something beyond your control, like drought (as did the couple in the vignette about Fortaleza, Brazil)? If so, what was this like; if not, what would this be like?
- What factors drew you to the place where you live now (pull factors)? Did you have to worry about whether there would be enough jobs, housing, and services, like many people do in Middle and South America?
- What ties do you still have to the place you once lived? Has your decision to live where you are now influenced any others to migrate to your location?
- Do you plan to move away from your current location, perhaps after graduation? If so, what are the push and pull factors that will influence your move?

2. Participation in the formal economy

By examining the transitions associated with NAFTA and other trade agreements in the region, we know that many people are taking jobs in the formal sector, leaving their livelihoods in the informal sector behind.

- List new areas of employment in which people in the region are participating (e.g., maquiladoras and the tourism industry).
- What may be the positive effects of this transition on one's life, as well as their family?
- What may be the negative effects of this transition on one's life, as well as their family?
- What might life be like in this region if people could only work in the formal economy?

3. Extended families

Extended families are very important and are the basic social institution in all societies in Middle and South America.

- What are some of the benefits of an extended family in this region? How does it affect daily life?
- Do you have an extended family? Do they live nearby or even in the same house? Why do you think this is so?
- What do you think the positive and negative effects of an extended family are or would be in your life?

4. Are *machismo* and *marianismo* present in your society?

Gender roles in Middle and South America are based heavily on the concepts of *machismo* and *marianismo.*

- Outline the characteristics and roles of *machismo* and *marianismo.*
- What are the pros and cons of the roles of *machismo* and *marianismo*? Consider their effects on both males and females.

- Do you see similar roles in the society in which you live? How are they similar and different? Why is this the case?
- Describe any changes you foresee in gender roles in the years to come in your society.
- How would changes such as these affect the development of Middle and South America?

5. Measures of human well-being

GDP ignores certain aspects of well-being that HDI and GEM take into account. Each measure, however, does give us important information about a place. Much can be revealed when we look at all three measures in one country.

- Using the table of human well-being rankings (Table 3.3), choose a country in this region that has data for all three rankings.
- Using the textbook *and* the library or Internet (start with the CIA World Factbook: www.cia.gov/cia/publications/factbook), research this country and attempt to explain what factors contribute to each of these three index values.
- Discuss any measures that seem to contradict each other (or are significantly different).

IMPORTANT PLACES

The following places are featured in the chapter. Make sure you can locate all of them on a map. Blank outline maps can be found on the textbook's Web site: www.whfreeman.com/pulsipher4e. Also, to prepare for quizzes and exams, write a few important facts about each place in the space provided.

Physical Features

1. Amazon Basin

2. Amazon River

3. Andes Mountains

4. Atacama Desert

5. Baja California

6. Brazilian Highlands

7. Caribbean Sea

8. Falkland Islands

9. Galapagos Islands

10. Greater Antilles

11. Guiana Highlands

12. Gulf of Mexico

13. Hispaniola

14. Lake Titicaca

15. Lesser Antilles

16. Orinoco River

17. Pampas

18. Parana River

19. Patagonia

20. Rio Grande River

21. Sierra Madre Occidental

22. Sierra Madre Oriental

23. Tierra del Fuego

24. Yucatan Peninsula

Regions/Countries/States/Provinces

25. Antigua and Barbuda

26. Argentina

27. Bahamas

28. Barbados

29. Belize

30. Bolivia

31. Brazil

32. Chile

33. Colombia

34. Costa Rica

35. Cuba

36. Dominica

37. Dominican Republic

38. Ecuador

39. El Salvador

40. French Guiana

41. Grenada

42. Guadaloupe

43. Guatemala

44. Guyana

45. Haiti

46. Honduras

47. Jamaica

48. Martinique

49. Mexico

50. Montserrat

51. Netherlands Antilles

52. Nicaragua

53. Panama

54. Paraguay

55. Peru

56. Puerto Rico

57. St. Kitts and Nevis

58. St. Lucia

59. St. Vincent and the Grenadines

60. Suriname

61. Trinidad and Tobago

62. Uruguay

63. Venezuela

Cities/Urban Areas

64. Asunción

65. Belmopan

66. Bogotá

67. Brasília

68. Buenos Aires

69. Cancun

70. Caracas

71. Cayenne

72. Chiapas

73. Curitiba

74. Fortaleza

75. Georgetown

76. Guatemala City

77. Havana

78. Kingston

79. La Paz

80. Lima

81. Managua

82. Mexico City

83. Montevideo

84. Panama City

85. Paramaribo

86. Port-au-Prince

87. Quito

88. Rio de Janeiro

89. San José

90. San Juan

91. San Salvador

92. Santiago

93. Santo Domingo

94. Sao Paulo

95. Sucre

96. Tegucigalpa

MAPPING EXERCISES

The following are three mapping exercises to improve your knowledge of the location of places, underscore why they are important, and clarify how they relate to each other. Some questions will ask you to locate places, compare maps, or fill in data; others will test your understanding of *why* you were asked to map the features that you did. Use the blank outline maps at the end of the chapter to complete these exercises. Additional blank outline maps can be found on the textbook's Web site: www.whfreeman.com/pulsipher4e.

1. Landforms, the environment, and population density

The region of Middle and South America extends south from the midlatitudes of the Northern Hemisphere, across the equator, nearly to Antarctica. Within this vast expanse, there is a wide variety of highland and lowland landforms; some landforms are more conducive to human settlement than others.

- Draw the general outline and label the following landforms from the regional map of Middle and South America (Figure 3.1): Sierra Madre Occidental, Sierra Madre Oriental, Guiana Highlands, Andes Mountains, Brazilian Highlands, Atacama Desert, and Patagonia.
- Also, trace the Amazon River with a thick blue line, and draw the outline of the Amazon Basin with a lighter blue line.
- Using the map of population density (Figure 3.12), shade the areas (in red) that have over 100 people per square kilometer (over 261 people per square mile).

Questions

a. Analyzing your map, describe the general pattern of population distribution.
b. Explain why some highland areas have high population concentrations. Explain why some highland areas have low population concentrations.
c. Explain why some lowland areas have high population concentrations. Explain why some lowland areas have low population concentrations.

2. Population growth, female literacy, and well-being

Overall, investment in education is not sufficient and health care is generally poor in the region. Populations in most countries are still growing rapidly, which can have dramatic effects on future quality of life.

- Shade the 10 countries featured in Figure 3.13 according to their rate of natural increase (1975-2003): yellow (0.0-1.0); orange (1.1-2.0); and red (2.1 and higher).
- Using the table of human well-being rankings (Table 3.3), write in the female literacy rate for all countries in the region.

- Use a graduated symbol (e.g., small to large circles) to map GDP per capita for all countries in the region. Use the following categories: $0-5,000; $5,001-10,000; and $10,001 and higher (Table 3.3).

Questions

a. Compare these population growth rates, female literacy rates, and GDP per capita to those for the United States. What does this tell you about the growth and quality of life in this region?
b. Analyzing the map, do you see any easily explainable patterns in the distribution of literacy? Why?
c. Draw some conclusions about the relationship between population growth and human well-being (using literacy rate and GDP as indicators of human well-being)? Are there any anomalies? If so, explain them.
d. How do you predict rapid population growth will affect this region in terms of GDP and female literacy?

3. Economy and well-being of Middle and South America

GDP per capita masks the very wide disparity of wealth in the region. Some HDI rankings are higher partly because education is somewhat more available across gender and class. GEM is low overall, but is higher in some countries because their governments support education and equal opportunity for women.

- Using the table of human well-being rankings (Table 3.3), shade countries from light blue to dark blue based on these categories of GDP per capita: $0-5,000; $5,001-10,000; and $10,001 and higher.
- Next draw hatch (///) patterns (most dense hatch for the highest ranking and widest hatch for the lowest ranking) over the shading based on these categories of HDI: 0-59; 60-119; and 120 and higher. Remember, high numbers are actually low rankings.
- Use a graduated symbol (e.g., from large to small squares: use larger squares for higher rankings) to map the GEM, using the following three categories: 0-27; 28-55; and 56 and higher. Again, remember that high numbers are actually low rankings.

Questions

a. Briefly describe the general relationship you would expect for the three variables you mapped.
b. Do you see any discrepancies between GDP and HDI (e.g., high GDP with low [high number] HDI or low GDP with high HDI)? Explain why these discrepancies may exist *in general*.
c. One might assume that a country with high GDP per capita would also have high (low number) HDI and GEM rankings. Which three countries stand out the most (i.e., have the most discrepancies)? Thoroughly explain the discrepancies for these specific countries based on history or current conditions.

SAMPLE EXAM QUESTIONS

The following are sample questions to help you review for an exam. Answers are found in the back of this workbook.

1. Despite much economic, political, and cultural diversity among the countries of Middle and South America, which one of the following characteristics do they have in common?
   A. Type of physical environment
   B. Experience as colonies of Europe
   C. Type of natural resource exploited for economic growth
   D. Ethnic composition of the population

2. The extremely dry climate of the Atacama Desert is the result of what two phenomena?
   A. The Peru Current and the Andes rain shadow
   B. El Niño and the Sierra Madre winds
   C. La Nina and the Amazon chinook
   D. The Patagonia Effect and the Magdalena tides

3. Which statement best describes the human landscape of Middle and South America just before its encounter with Europeans?
   A. Human settlements were found only on the eastern coast of what is now Brazil.
   B. The region was nearly devoid of humans except for small groups of nomadic peoples.
   C. Many people were living in socially organized, technologically advanced cities.
   D. The Incan Empire, concentrated in the highlands of Middle America, was the only human settlement at the time.

4. Which of the following characterizes the regional trend in the rate of natural increase?
   A. Slowly increasing but still low
   B. Rapidly declining but remains high
   C. Slowly declining but remains high
   D. Rapidly increasing but still low

5. Which of the following describes the principal migration trend in Middle and South America?
   A. North-to-south movements
   B. Rural-to-urban movements
   C. Island-to-mainland movements
   D. Coastal-to-interior movements

6. Which of the following represents a legacy of colonialism in Middle and South America?
   A. The relative wealth of national governments
   B. A manufacturing-supportive infrastructure
   C. The region's dependence on extraction-based activities
   D. Financial networks that allow local growth from domestic profits

7. Why do many governments in Middle and South America favor export-oriented agriculture?
   A. It decreases the flow of profits from the region.
   B. It brings in large amounts of cash with which they can repay debt.
   C. It improves the supply of food for local consumption.
   D. It is an effective means of smuggling illegal drugs into other countries.

8. Which of the following has NOT occurred with the introduction of structural adjustment programs in Middle and South America?
   A. Health care spending has declined.
   B. Public spending on schools has decreased.
   C. National debt burdens have been reduced.
   D. The number of people living in poverty has decreased.

9. Which of the following did the textbook report as a criticism of the United States' "war on drugs"?
   A. It has focused too much on users and traffickers in the United States and not on the producing countries in Middle and South America.
   B. It focuses too much on Asian countries despite higher rates of drug production in Middle and South America.
   C. U.S. officials have been paid by Middle and South American officials to end the "war on drugs."
   D. It has greatly increased the profile of the U.S. military in many countries in Middle and South America.

10. Which of the following is a principal contributor to the decline of forests in the Amazon Basin?
    A. Multinational logging companies
    B. An increase in small private farms
    C. Global climate change
    D. An increase in large automobile manufacturing plants

MIDDLE & S. AMERICA
0 Miles 1000
0 Kilometers 1600

MIDDLE & S. AMERICA
0 Miles 1000
0 Kilometers 1600

MIDDLE & S. AMERICA
0
Miles
1000
0
Kilometers
1600

CHAPTER FOUR
Europe

LEARNING OBJECTIVES

After reading the chapter and working through this study guide, you should:

- Know why the climate in Europe is reasonably mild despite its northerly latitude. Know how agriculture has adapted to the variations in climate in each subregion.
- Know that Europe has a high population density living primarily in urban settings with a very high standard of living, despite a diminishing resource base.
- Know how the supranational organization, the European Union (EU), promotes economic, political, and related social integration, all to eliminate regional disparities and provide a high standard of living in all member countries.
- Know that guest workers (temporary migrants) from Turkey, North Africa, and the poorer parts of Europe perform many of the region's menial jobs. Understand that they often face difficulties with acculturation and assimilation into majority cultures, which themselves are seeking to maintain national identities in the face of the homogenization associated with EU membership.
- Know that while gender roles are gradually changing, women must still cope with challenges of day care, flexible working hours, gendered work roles, and home responsibilities when they work outside the home.
- Know the differences among varying welfare programs provided by European democratic governments.
- Know that environmentalists in Europe are focusing on sustaining livable environments and encouraging regional cooperation in dealing with pollution in the seas, rivers, and air.

KEY TERMS

The following terms are in bold in the textbook. In the space next to the term (or on a separate sheet or flash cards), you can fill in the definitions for reference or quiz yourself for exam review. Definitions are found in the glossary of the textbook.

acculturation

Age of Exploration

assimilation

basin

centrally planned

cold war

Council of the European Union

cultural homogenization

democratic institutions

devolution

double day

dumping

economic integration

economies of scale

ethnic cleansing

euro

European Commission

European Economic Community (EEC)

European Parliament

European Union (EU)

exclave

feudalism

Green

guest workers

Holocaust

humanism

humid continental climate

iron curtain

medieval period

Mediterranean climate

mercantilism

nationalism

North Atlantic Treaty Organization (NATO)

producer services

Protestant Reformation

Renaissance

resource base

Roma

Scandinavia

Schengen Accord

social welfare

subsidies

temperate midlatitude climate

welfare state

world cities

REVIEW QUESTIONS

The following review questions are related directly to the textbook material. These questions can be used to help you prepare for an exam or you may want to read through the questions before you begin reading the textbook, quizzing yourself after you complete each section.

The Geographic Setting

1. What is meant by the description of the region as "peninsulas upon peninsulas"? What is the climatic impact of such extensive contact with oceans and seas?
2. Of the four types of landforms in this region (mountains, uplands, lowlands, and rivers), which are most closely associated with current human activities? What are some of these activities and where is the human impact most profound?
3. What unique adaptations have humans made to the three main climate zones (temperate midlatitude, Mediterranean, and humid continental)?
4. How were the Industrial Revolution and the process of colonialism linked?
5. Despite a diminishing natural resource base, Europe has maintained a high standard of living with promises that this will continue. How has the region managed to do so?
6. What are the historical economic influences that gave rise to the European Union (EU) and support its continued expansion? What remains as the central reason for the establishment of the EU?
7. Even though Europe has a very high population density, most people enjoy a high standard of living. What aspects of Europe's urbanization allow for a quality of life amidst high population densities?
8. What burdens are placed on a society when the population of a country achieves a negative rate of natural increase? What are some coping mechanisms?

Current Geographic Issues

9. Although the EU began as an economic partnership, other issues have come to the forefront as equally important issues for the EU to address. Identify these issues and predict their impact on the EU members in the twenty-first century.
10. What are some policies and practices employed by the EU to help maintain the global competitiveness of the region's economy?
11. From the perspective of individual countries, what are some negative or more challenging aspects of attaining and maintaining membership in the EU?
12. Identify two countries that elected not to use the euro. What rationale did these countries provide for retaining their own currency? What are the disadvantages?
13. Central European countries comprise the largest bloc of new countries gaining membership in the EU. What are the advantages to these countries for joining the EU? What are the disadvantages they will have to overcome?
14. The growth of the service industry is associated with increased employment opportunities. What are some of these opportunities? What is the overall mixture of high paying vs. low-paying jobs?
15. How do European countries respond to the globalization of the economy in the context of agriculture?
16. What trends do the authors suggest are occurring because of more open borders and changing perspectives on citizenship requirements for immigrants?
17. Despite many enlightened social attitudes, this region is conservative in its attitudes toward gender roles in society. What accounts for these pervasive attitudes, and

how do they affect women's job opportunities and home responsibilities?

18. Link the role of women as workers, caregivers, and providers in the context of the differing welfare systems in Europe.
19. If Europe's natural landscape has been so extensively transformed by humans over time, then what particular role can the Green movement play? What do you see in the Green movement that may serve as important role models for the rest of the world?
20. Describe the physical geography characteristics and the social, economic, and political issues that may explain the high rates of pollution in the Black, Baltic, and Mediterranean Seas.

CRITICAL THINKING EXERCISES

The following are "what if"-type questions that illustrate concepts you are learning from the textbook. These questions ask you to apply the ideas and principles you learned from the textbook to new situations.

1. Marginalized groups in European countries

After reading about the daily lives of so many Europeans, identify one group of people that you think faces the most challenges in improving their status of human well-being. Your choice may be based on a theme such as ethnicity, gender, age, religion, or nationality. The subregion section of the text may be useful for this exercise.

- Explain why you selected this particular group.
- Identify the issues or policies that are marginalizing this particular group. What is the kind of marginalization that this group is experiencing?
- Suggest at least three ways to improve the lives of this particular group.
- Suggest challenges that will be faced in bringing about the change.

2. Europe's negative rate of natural increase: How do you fit in?

Many European countries are experiencing a negative rate of natural increase. Soon a significant percent of the population will be aging.

- Select a European country that is currently experiencing a negative rate of population growth. Suppose that you are a citizen of that country. Use the CIA World Factbook at www.cia.gov/cia/publications/factbook for necessary data and information.
- Find the population pyramid for that country on the U.S. Census Bureau's Web site: www.census.gov/ipc/www/idbpyr.html.
- Identify where you would fit into that country's population pyramid.
- Suggest responsibilities that people your age might be facing as a consequence of a growing population of elderly in your selected country.

3. Urban life in Europe

Europe is a highly urbanized region. Tourists from around the world are attracted by the art, music, and architecture of the great cities of this region. Assume that you have an opportunity to study abroad in a European city for one semester.

- Identify one city where you would like to study and specify the university that you would attend. Explain why you selected this university, its city, and its country.
- Using information from the text and visits to Web sites for the country, the city, and the university you have selected, explain how five examples of urban living in your selected European city and university would be different from the lifestyle you now experience as a student.
- Explain how your examples match the textbook's discussion of urbanization in Europe.

4. Supranational transitions in Europe: Challenges of integration

With the ever-increasing role of the European Union in the daily lives of Europeans, one wonders how citizens of individual countries are responding to such regional changes as the euro, a common passport, open borders, and the influence of a homogenous Euro-pop culture. Reflect on your own personal sense of nationalism and decide if you could support a United States of North America that included Canada, the United States, and Mexico.

- Identify the challenges you would face in developing loyalty to a supranational political structure.

5. Impact of EU membership on a selected country

Select a country that joined the EU in the twenty-first century. The official EU Web site (www.europa.eu) will help you with this information. You may want to investigate the country's official Web site as well. Respond to the following questions:

- What are the positive and negative aspects of EU membership for the country you selected?
- How has membership in the EU fostered greater linkages within the EU and the rest of the globe?
- What major problems is the country facing and how can its EU membership help it to solve those problems?

IMPORTANT PLACES

The following places are featured in the chapter. Make sure you can locate all of them on a map. Blank outline maps can be found on the textbook's Web site: www.whfreeman.com/pulsipher4e. Also, to prepare for quizzes and exams, write a few important facts about each place in the space provided.

Physical Features

1. Adriatic Sea

2. Alps

3. Arctic Ocean

4. Atlantic Ocean

5. Baltic Sea

6. Black Sea

7. Danube River

8. Elbe River

9. English Channel

10. Faroe Islands

11. French Riviera

12. Iberian Peninsula

13. Mediterranean Sea

14. Morava River

15. North European Plain

16. North Sea

17. Pyrenees

18. Rhine River and Delta

19. Sardinia

20. Sicily

21. Straits of Gibraltar

22. Svalbard

Regions/Countries/States/Provinces

23. Albania
24. Austria
25. Balkans
26. Baltic States
27. Basque country
28. Belgium
29. Benelux
30. Bosnia-Herzegovina
31. British Isles
32. Bulgaria
33. Catalonia
34. Central Europe
35. Champagne
36. Croatia
37. Cyprus
38. Czech Republic
39. Denmark
40. England
41. Estonia
42. Finland
43. France
44. Galicia

45. Germany

46. Great Britain (Britain)

47. Greece

48. Greenland

49. Hungary

50. Iceland

51. Ireland (Republic of)

52. Italy

53. Kaliningrad

54. Kosovo

55. Latvia

56. Lithuania

57. Luxembourg

58. Macedonia

59. Malta

60. Montenegro

61. Netherlands

62. Northern Ireland

63. North Europe

64. Norway

65. Poland

66. Portugal

67. Romania

68. Scandinavia

69. Scotland

70. Serbia

71. Silesia

72. Slovakia

73. Slovenia

74. South Europe

75. Spain

76. Sweden

77. Switzerland

78. Ukraine

79. United Kingdom

80. Upper Silesia

81. Vatican

82. Wales

83. West Europe

Cities/Urban Areas

84. Amsterdam

85. Antwerp

86. Athens

87. Augsburg

88. Avignon

89. Barcelona

90. Belfast

91. Belgrade

92. Berlin

93. Bern

94. Bratislava

95. Bruges

96. Brussels

97. Bucharest

98. Buchenwald

99. Budapest

100. Cambridge

101. Copenhagen

102. Dijon

103. Dresden

104. Dublin

105. Florence

106. Frankfurt

107. Genoa

108. (The) Hague

109. Helsinki

110. Innsbruck

111. Kohtla-Jarve (Latvia)

112. Krakow

113. Lisbon

114. Liverpool

115. Ljubljana

116. London

117. Luxembourg (City)

118. Madrid

119. Manchester

120. Marseille

121. Milan

122. Nicosia

123. Oslo

124. Paris

125. Prague

126. Reykjavik

127. Riga

128. Rome

129. Rotterdam

130. Sarajevo

131. Skopje

132. Sofia

133. Stockholm

134. Stuttgart

135. Tallinn

136. Tirane

137. Titograd

138. Valletta

139. Venice

140. Vienna

141. Vilnius

142. Warsaw

143. Zagreb

MAPPING EXERCISES

The following are three mapping exercises to improve your knowledge of the location of places, underscore why they are important, and clarify how they relate to each other. Some questions will ask you to locate places, compare maps, or fill in data; others will test your understanding of *why* you were asked to map the features that you did. Use the blank outline maps at the end of the chapter to complete these exercises. Additional blank outline maps can be found on the textbook's Web site: www.whfreeman.com/pulsipher4e.

1. River basins, population, and pollution

Rivers play an important role in the continued success of European economies. Significant rivers flow through several countries and provide valuable transportation access. Manufacturing districts grew up in areas that formerly provided important industrial materials. Unfortunately, these valuable rivers are also highly polluted. Using the map of Europe, complete the following:

- Label the following rivers: the Rhine and the Danube.
- Select a pattern to show the high-tech manufacturing and high quality/luxury goods manufacturing areas (Figure 4.20).
- Using the map of population density (Figure 4.11), shade the areas (in red) that have a population density of greater than 250 persons per square kilometer (greater than 650 persons per square mile).

- Locate and label all cities with 2 million people or more (Figure 4.11).
- Select a pattern to show the areas of high human impact on the land (Figure 4.33).

Questions

a. Using this mapped information, briefly describe the pattern and distribution of population.
b. Is population concentrated in any particular location (e.g., along the coasts, along major rivers, or in the interior)?
c. Is population concentrated near or away from manufacturing centers?
d. Summarize the relationship between the two rivers, population concentrations, manufacturing, and pollution.
e. Suggest some changes that will still allow the rivers to be used as major arteries for transport while at the same time permitting successful cleanup and good stewardship.

2. De-industrialization of Europe: High-tech manufacturing in the twenty-first century.

Figure 4.20 shows the geographic change of the industrial centers in Europe as high-tech production has become a major economic activity. Figure 4.3 (map of the European Union) lists the dates of each country's entry into the EU, and Figure 4.21 depicts the transportation network for this region. After a careful reading of the section on the EU and observation of the three maps, respond to the following questions.

a. How does the growth of the EU correlate with the shift to high-tech manufacturing?
b. What is the relationship between the transportation network and the growth of the high-tech centers?
c. To what extent is the EU policy of extending development into previously marginalized areas reflected in the maps showing centers of production?
d. What efforts by the EU are needed to extend the high tech production into even more of the marginalized areas? Give some examples of good practices that could benefit some of the newest EU members.

3. Women's empowerment in Europe

Some European countries have shown exemplary progress in bringing women into equally responsible positions in government and related socio-economic strata of society. However, these countries may not be representative of the entire European region.

- On the blank map of Europe, map GEM as listed in Table 4.1 (human well-being rankings) using graduated circles for each country.
- Map the welfare systems (Figure 4.32) using a different color for each of the different welfare systems.

Questions

a. Describe any patterns that suggest a correspondence between certain welfare regimes and women's overall empowerment in different European countries.

b. Based on your reading of the textbook, what do you think is the driving force in areas where women are highly empowered and actively engaged in political agendas?

c. After reading the textbook sections on gender and social welfare systems (along with reference to the section on subregions), suggest a country in Europe that could serve as a good role model for promoting women's empowerment. Give three reasons for why you selected this country.

SAMPLE EXAM QUESTIONS

The following are sample questions to help you review for an exam. Answers are found in the back of this workbook.

1. What is true of Paris and London?
   A. They are, although well known, relatively small with populations under 1 million each.
   B. They are unimportant on the global scale.
   C. They are world cities of cultural and economic significance.
   D. They have been of global importance since Roman times.

2. What movement arose to challenge the elitist practices of the Catholic Church?
   A. Liberation theology
   B. Anglican resurgence
   C. Methodist renaissance
   D. Protestant reformation

3. The creation of the EU formalizes which of the following processes?
   A. Economic integration of Europe
   B. Rising self-sufficiency of European countries
   C. Spread of European influence into northern Africa
   D. The replacement of NATO by a similar European-led alliance

4. Which of the following statements reflects the concept of the *double-day?*
   A. A male-female couple can double the household's daily income if they both work outside the home.
   B. Due to low wage employment, immigrants in Europe typically work two consecutive shifts in one day to support their families.
   C. Women who work outside the home still perform the largest share of domestic work.
   D. Through high-tech mechanization, European firms have been able to double their daily labor productivity.

5. Which of the following is largely responsible for the warm and wet climate of northwestern Europe?
   A. The North Atlantic drift
   B. The intertropical convergence zone
   C. The Scandinavian chinook
   D. The North African stream

6. Compared to the North and Atlantic seas, with what problem must the Mediterranean, Baltic, and Black seas contend?
   A. They have difficulty cleansing themselves of pollutants due to their location.
   B. Their substantial oil and gas resources are being developed recklessly.
   C. Each is shrinking because the main rivers that feed them have been diverted.
   D. Freshwater animal life is being threatened by saltwater intrusion.

7. What was the dominant type of housing constructed to accommodate the post-World War II growth of Europe's cities and is the type of lodging in which most people now reside?
   A. Detached single-family homes
   B. Extended-family compounds
   C. Apartments
   D. Communal dormitories

8. What does "acculturation" mean?
   A. Comprehensive, total change of lifestyle to that of a host country
   B. Adaptation to a host culture to the extent needed to function effectively and be self-supporting
   C. Invention of a totally new culture as a result of a new environment
   D. Discarding of culture entirely for a non-cultural existence

9. Which is NOT true of Europe's population trends?
   A. Fewer children are being born than in the 1970s.
   B. Thirty-five to 40 percent of people have no children at all.
   C. More children are being born to unmarried parents.
   D. The birth rate is on the rise.

10. Which of the following constitutes the major source of population growth in Europe?
   A. Increasing birth rates
   B. Immigration
   C. Rapidly falling death rates
   D. Decreasing infant mortality

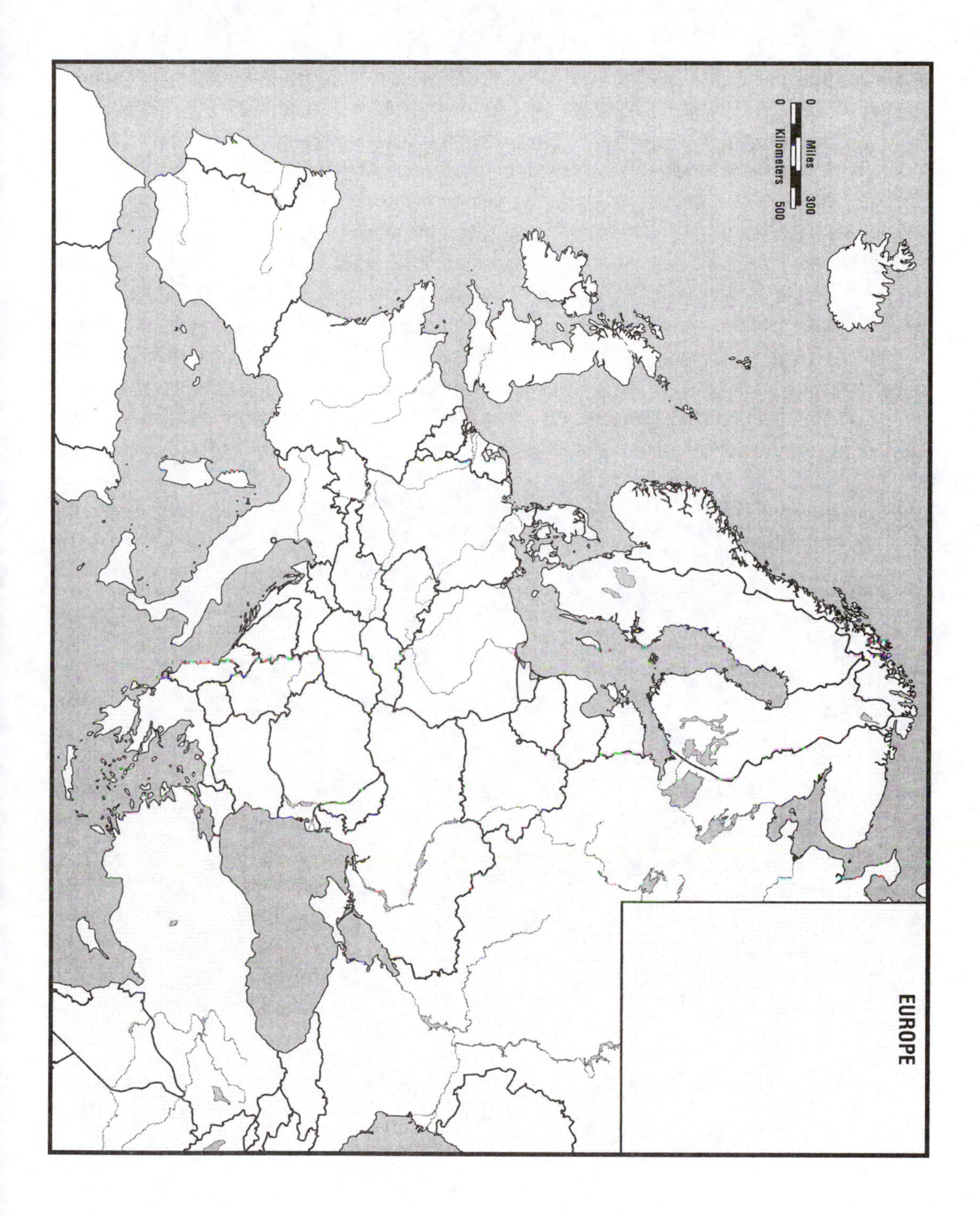

EUROPE
0 Miles 300
0 Kilometers 500

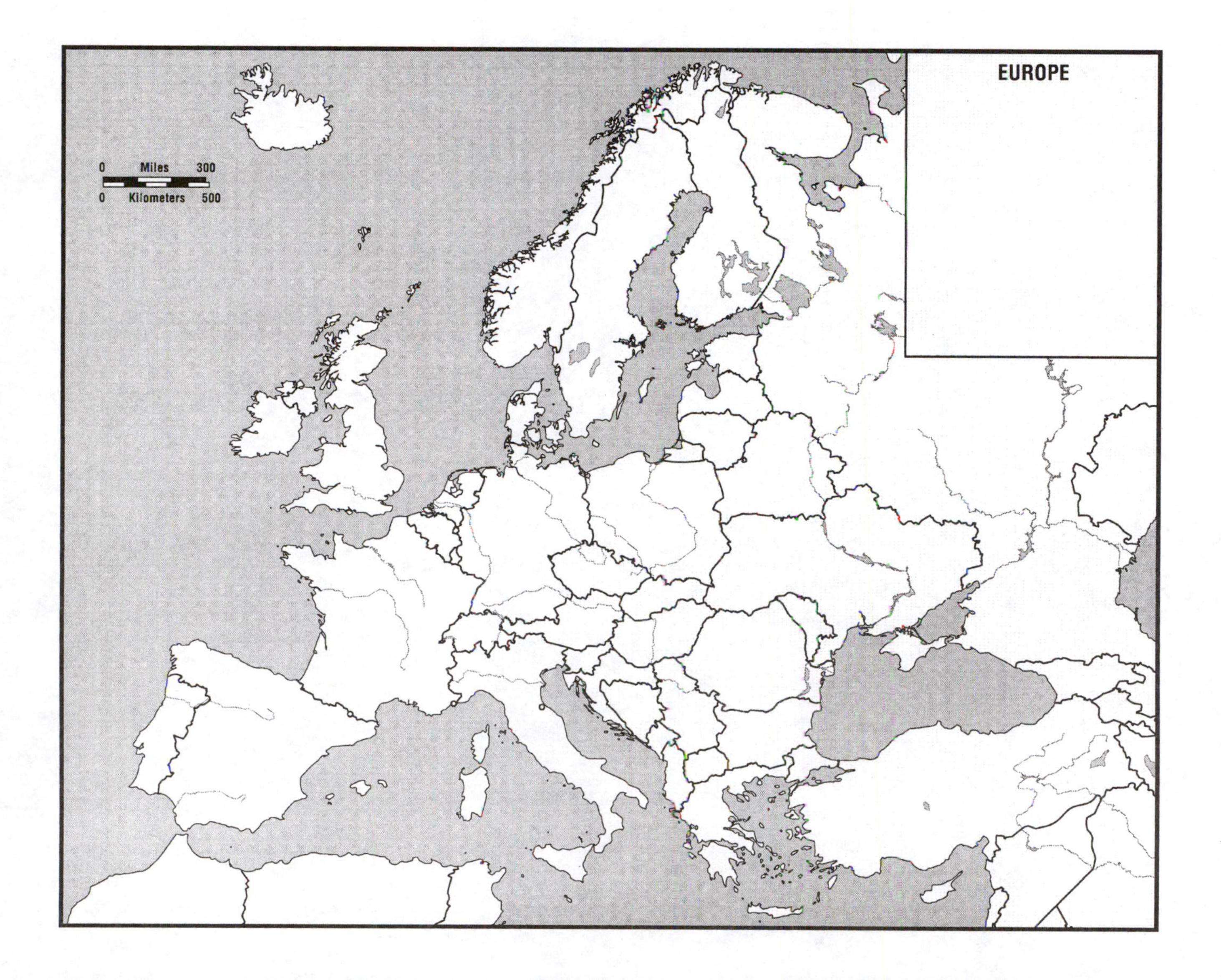
EUROPE
0 Miles 300
0 Kilometers 500

CHAPTER FIVE
Russia and the Newly Independent States

LEARNING OBJECTIVES

After reading the chapter and working through this study guide, you should:

- Know why the years since the breakup of the Soviet Union have been a time of turmoil and transition, as 15 independent republics attempt to make the transition to democratic governments and embark on market economies.
- Understand the challenges of living in the different subregions, each with unique physical characteristics, but all sharing a continental climate.
- Know the reasons for demographic changes in this region, especially the decline in birth rates and the decline in life expectancy.
- Understand how Russia and other states are using their abundant natural resources to help move their economies forward.
- Understand the challenges of developing democratic systems of governance, especially in helping people to accept civic responsibilities. Also understand the challenge of confronting corruption, which is becoming endemic in many sectors.
- Know how the transition to a market economy has led to special difficulties for women and the aging.
- Know how the exploitation of resources has damaged the region's environment. Understand how the extent of this damage affects quality of life.

KEY TERMS

The following terms are in **bold** in the textbook. Students are given space in the workbook to fill in the definitions for reference or quiz themselves for exam review. Definitions are found in the glossary of the textbook.

Bolsheviks

capitalists

centrally planned economy

command economy

Communism

Communist party

czar

Gazprom

glasnost

Group of Eight (G8)

marketization

Mongols

nomadic pastoralists

nonpoint sources of pollution

oligarchs

perestroika

permafrost

privatization

Russian Federation

Russification

serfs

Slavs

socialism

Soviet Union

steppes

taiga

tundra

underemployment

Union of Soviet Socialist Republics (USSR)

REVIEW QUESTIONS

The following review questions are related directly to the textbook material. These questions can be used to help you prepare for an exam or you may want to read through the questions before you begin reading the textbook, quizzing yourself after you complete each section.

The Geographic Setting

1. Discuss the climatic conditions that make European Russia and Caucasia the agricultural backbone of the region.
2. Russia is called the epitome of a continental climate. What are two characteristics of that continentality? List three impacts on human activity in such a climate.
3. The Russian-Mongol interaction was historically one of domination and exchange. How did both sides benefit from mutual cooperation?
4. This region claims to be part of the developed world, yet some of the population patterns discussed contradict this assertion. What are three of these contradictory facts and what circumstances gave rise to them?
5. Discuss the surprising contradiction that command economies are at the same time faced with massive gluts and acute scarcities.
6. How is economic change since the dissolution of the USSR reflected in household well-being? What are short-term and long-term challenges faced by households?
7. What are some of the reasons why male life spans in the Russian Federation are now the shortest of those in the industrialized countries?

Current Geographic Issues

8. Why do citizens of this region have to formally learn about democracy and capitalism? Why are these particularly difficult lessons? Identify any groups that appear to be more adept at learning these two themes and tell why they are successful.
9. How is the command economy an explanation for the lack of management and technical innovation during the Soviet era?
10. What is the basis for Russians' well-developed skill at tax dodging? Name two repercussions on society as a whole.
11. What are some reasons that the Russian economy has a much lower rate of foreign investment than other economies undergoing the transition from centrally planned to capitalistic systems (e.g., Eastern Europe and China)?
12. What role have the natural gas and oil reserves begun to play in the economies of these countries?
13. How was Russia's expansion to the south and the east typical of colonialism? What has been the long-term impact on the colonized areas?
14. Why is the military such an important issue for the leaders in Moscow?
15. How does the loss of jobs in the government-owned industries also represent the loss of a critical social safety net?

16. In what ways are the aging members of these states particularly affected by economic reforms?
17. How are the universal stereotypical roles of women reinforced in the current economic situation in Russia? How is this detrimental to women's roles in society and at home?
18. How is the region's environmental plight rooted in Communist drives to industrialize the USSR?
19. What did the Russian environmentalist mean when he said, "When people became more involved with their stomachs, they forgot about ecology"?
20. What strategies did this region develop in order to survive the economic transitions to a market economy? Which ones are the most successful in the long term?

CRITICAL THINKING EXERCISES

The following are "what if"-type questions that illustrate concepts you are learning from the textbook. These questions ask you to apply the ideas and principles you learned from the textbook to new situations.

1. Life in Russia for a university student

After reading the sections in the text that discuss the well-being and opportunities for the Russian population, you may have thought about how you would or would not cope with comparable circumstances.

- Place yourself somewhere in Russia and consider the quality of life you might have, as well as your prospects for the future, if you were a student there.
- Where would you live and what kinds of work or internships would you most likely find upon completing your education?
- Identify the impacts of massive pollution on you as a college student. Identify the impacts of current economic reforms on you based on your gender and age. What would be your future prospects for economic security and good health in this society?

2. Growing old in societies under transition

During the Soviet era, adequate retirements were guaranteed, and workers had expectations of reasonable pensions during their retirement years. With the transition to market economies and the decline of state-owned corporations, the elderly in these countries are struggling to survive.

- Investigate how the elderly who worked in former state-owned companies are managing to survive in one of the countries of this region.
- Identify the efforts on the part of central government to provide for these marginalized members of society.
- Identify how the current workers in privatized companies are planning for their own retirements, and compare the new system to the old.

3. Humans can change climate!

The desertification that is occurring around the Aral Sea is a classic case study of human-induced deserts (also called anthropological desertification).

- What are the policies and actions followed at the national and subregional levels that have brought on this climate change?
- What are some steps that could be followed to reverse the process while providing sustainable options for economic livelihoods? Suggest some roles that citizens can have in this process.

4. How did you learn about capitalism? How does it compare to the process of learning about capitalism in this region?

Your experience with capitalism has most likely occurred over a long period of time and usually outside the context of formal education.

- What is the process by which you learned about capitalism and democracy?
- Compare and contrast this process to the one that citizens in Russia and the new republics are experiencing as they learn about capitalism, the functioning of free-market economies, and democracy.

5. Individual efforts to enhance nutrition

The textbook includes an informative note on urban gardening in this region. Even during the Soviet era, citizens were allowed to have gardens in small plots, even to sell some of the produce in local open-air markets.

- Investigate the history of the *dacha* in Russia. Then investigate community garden projects in the United States.
- What do you see as similar, especially the reasons for participation in urban gardening?
- What are some differences between the urban gardens of Russia and the community gardens in the U.S. cities? Why do these differences exist?
- What would be a factor to motivate you to participate in a community garden and would that be similar to those motivating people in Russia who practice urban gardening? What would be some differences?

IMPORTANT PLACES

The following places are featured in the chapter. Make sure you can locate all of them on a map. Blank outline maps can be found on the textbook's Web site: www.whfreeman.com/pulsipher4e. Also, to prepare for quizzes and exams, write a few important facts about each place in the space provided.

Physical Features

1. Altai Mountains

2. Amu Darya River

3. Amur River

4. Aral Sea

5. Arctic Ocean

6. Baltic Sea

7. Black Sea

8. Carpathian Mountains

9. Caspian Sea

10. Caucasus Mountains

11. Central Siberian Plateau

12. Gulf of Finland

13. Hindu Kush Mountains

14. Kamchatka Peninsula

15. Kolmya River

16. Lake Baikal

17. North European Plain

18. Ob River

19. Pacific Ocean

20. Pamir Mountains

21. Sakalin Island

22. Sea of Okhotsk

23. Siberia

24. Sikhote Alin

25. Syr Darya River

26. Tien Shan Mountains

27. Transcaucasia

28. Ural Mountains

29. Volga River

30. West Siberian Plain

31. White Sea

Regions/Countries/States/Provinces

32. Abkhazia

33. Armenia

34. Azerbaijan

35. Belarus

36. Caucasia (Caucasian Republics)

37. Central Asia

38. Chechnya

39. Dagestan

40. European Russia

41. Georgia

42. Ingushetiya

43. Kaliningrad

44. Kazakhstan

45. Kyrgyzstan

46. Moldova

47. Nagorno Karabakh

48. Russia

49. Tajikistan

50. Tatarstan

51. Turkmenistan

52. Tuva

53. Ukraine

54. Uzbekistan

Cities/Urban Areas

55. Angarsk

56. Ashkhabad

57. Astana

58. Baku

59. Bishkek

60. Bukhara

61. Chelyabinsk

62. Chernobyl

63. Chisinau

64. Donetsk

65. Dushanbe

66. Grozny

67. Irkutsk

68. Kazan

69. Kiev

70. Krasnoyarsk

71. Minsk

72. Moscow

73. Nakhodka

74. Norilsk

75. Novosibirsk

76. Odessa

77. Samarkand

78. Sevensk

79. St. Petersburg

80. Tashkent

81. Tbilisi

82. Vladivostok

83. Volgograd

84. Yekaterinburg

85. Yerevan

MAPPING EXERCISES

The following are three mapping exercises to improve your knowledge of the location of places, underscore why they are important, and clarify how they relate to each other. Some questions will ask you to locate places, compare maps, or fill in data; others will test your understanding of *why* you were asked to map the features that you did. Use the blank outline maps at the end of the chapter to complete these exercises. Additional blank outline maps can be found on the textbook's Web site: www.whfreeman.com/pulsipher4e.

1. Industry, pollution, and populations at risk

We can learn much about the populations at risk in this region by using maps to generate information.

- Construct a map that shows areas with population densities of 261 people or more per square mile (101 people or more per square kilometer) (Figure 5.13). A good symbolization would be shading using the darkest shade to represent highest population concentrations.
- Overlay areas of high industrial development (Figure 5.20). Categorize these by type of industrial region as designated in the legend for Figure 5.20.
- Overlay areas of oil and natural gas producing regions (Figure 5.21). Again use a unique symbol, such as cross-hatching.
- Finally, overlay areas of human impacts (Figure 5.30) using unique symbols for high impact and medium-high impact.

Questions

a. What appears to be a major source of employment for people in the densely populated areas?
b. What conclusions can you draw about environmental impacts on health? What are the governmental responsibilities for citizens' health?
c. What are three ways this region can cope with the environmental impacts of industrialization, resource extraction, and production?

2. Population issues in Russia and the newly independent states

By mapping demographic data by country, you should be able to arrive at some conclusions about population growth in this region.

- Using the CIA World Factbook at www.cia.gov/cia/publications/factbook, construct a map that portrays the birth rate for each country in this region. Use graduated circles to represent birth rates.
- Refer to Table 5.2 (human well-being rankings) to construct an overlay that depicts GDP per capita for each country. Be sure the lightest shade represents the lowest value and the darkest shade represents the highest value.
- Refer to Table 5.2 to construct another overlay that depicts the HDI for each country. Use diagonal lines (///) to indicate HDI. Use closely spaced lines to represent those countries with good HDI values (low numbers) and use widely spaced lines to represent those countries with poor (high number) HDI values.

Questions

a. Use this mapped information to help explain why some countries have a decreasing birth rate and why others have an increasing birth rate.
b. If GDP and HDI are not fully explanatory for some of the countries, suggest three other factors that might explain those countries' birth rates.

3. Water and oil: Critical resources that affect international relations in the Central Asian Republics

Control and access to headwaters or upstream waters of the major rivers of the Central Asian Republics may be a deciding factor in the continuation of cotton production in the region. Oil pipeline routes may change as newly independent countries develop international ties with countries that were never part of the former USSR.

- Label the Central Asian Republics and label their neighbors.
- Use the regional map (Figure 5.1) to label and outline the rivers (in blue) that supply water to Central Asia's irrigation projects. Shade in all the countries that are drained by these rivers.
- On Figure 5.7 (map of agriculture), observe the cash crops in the Central Asia valleys. These are most likely cotton crops. Shade these areas on your map.
- Use the oil and natural gas resource map (Figure 5.21) to cross-hatch the location of oil- and gas-producing areas in this region.
- Also from Figure 5.21, draw in red the current oil and gas pipelines that move these resources to shipping ports. Using a second color, suggest alternative routes that lie outside of the region for moving this oil to shipping ports.

Questions

a. Identify countries that will have to develop some basis for negotiating use of the water from these rivers. How will upstream countries have an advantage? What counterarguments or possible offers can the downstream countries (where the rivers drain last) provide in order to assure access to the water?

b. Make an argument for how the new pipelines you suggested would be the most plausible considering international relations in the region, international demand, security, and certainly environmental sensitivity.

SAMPLE EXAM QUESTIONS

The following are sample questions to help you review for an exam. Answers are found in the back of this workbook.

1. Which one of the following features on the Eurasian physical landscape separates European Russia and Western Siberia?
   A. Volga River
   B. Ural Mountains
   C. Aral Sea
   D. Ob River

2. Which one of the following describes Russia's continental climate?
   A. Cool summers and warm winters
   B. Hot summers and cold winters
   C. Hot summers and warm winters
   D. Cold summers and cold winters

3. What type is the dominant groundcover in Russia's *steppe* lands?
   A. Coniferous forest
   B. Desert scrub
   C. Grasses
   D. Temperate rain forest

4. Which of the following is NOT one of the main efforts of Europe and the United States aimed at bringing Russia into closer formal association with established trading institutions?
   A. Inviting Russia to join the World Trade Organization
   B. Inviting Russia to increase its participation in the meetings of the Group of Eight
   C. Encouraging Russia to work toward conformity with European standards by admitting former Russian allies to the European Union (EU)
   D. Inviting Russia to join the EU

5. Which of the following describes the primary spatial growth of the Russian Empire beginning in the 1500s?
   A. From Mongolia to the north and west
   B. From Kazan to the north and west
   C. From Moscow to the south and east
   D. From Petersburg to the south and west

6. In the Soviet Union, what institution provided most of one's basic social services?
   A. Nongovernmental organizations
   B. Work place
   C. Extended family
   D. Military

7. Which of the following is cited in the textbook as a major reason for falling birth rates during the 1990s in Russia?
   A. The Russian government implemented population control policies.
   B. Couples decided against having children because of bleak economic prospects.
   C. There has been a significant increase in migration to cities where children are more of an economic liability.
   D. High levels of environmental pollution have caused sterility in many women.

8. The environmental and climatic changes wrought by the diversion of water from the Syr Darya and Amu Darya rivers are reflected in which of the following phenomena?
   A. Saltwater intrusion in the Black Sea
   B. Shrinkage of the Aral Sea
   C. Heavier and more frequent rains in Central Asia
   D. Cooler average temperatures throughout Central Asia

9. In the Soviet Union, what type of economy governed the production, distribution, and consumption of goods?
   A. Market
   B. Feudal
   C. Mercantile
   D. Command

10. In which of the following subregions do most of Russia's citizens reside?
    A. European Russia
    B. Western Siberia
    C. Central Siberia
    D. Far East

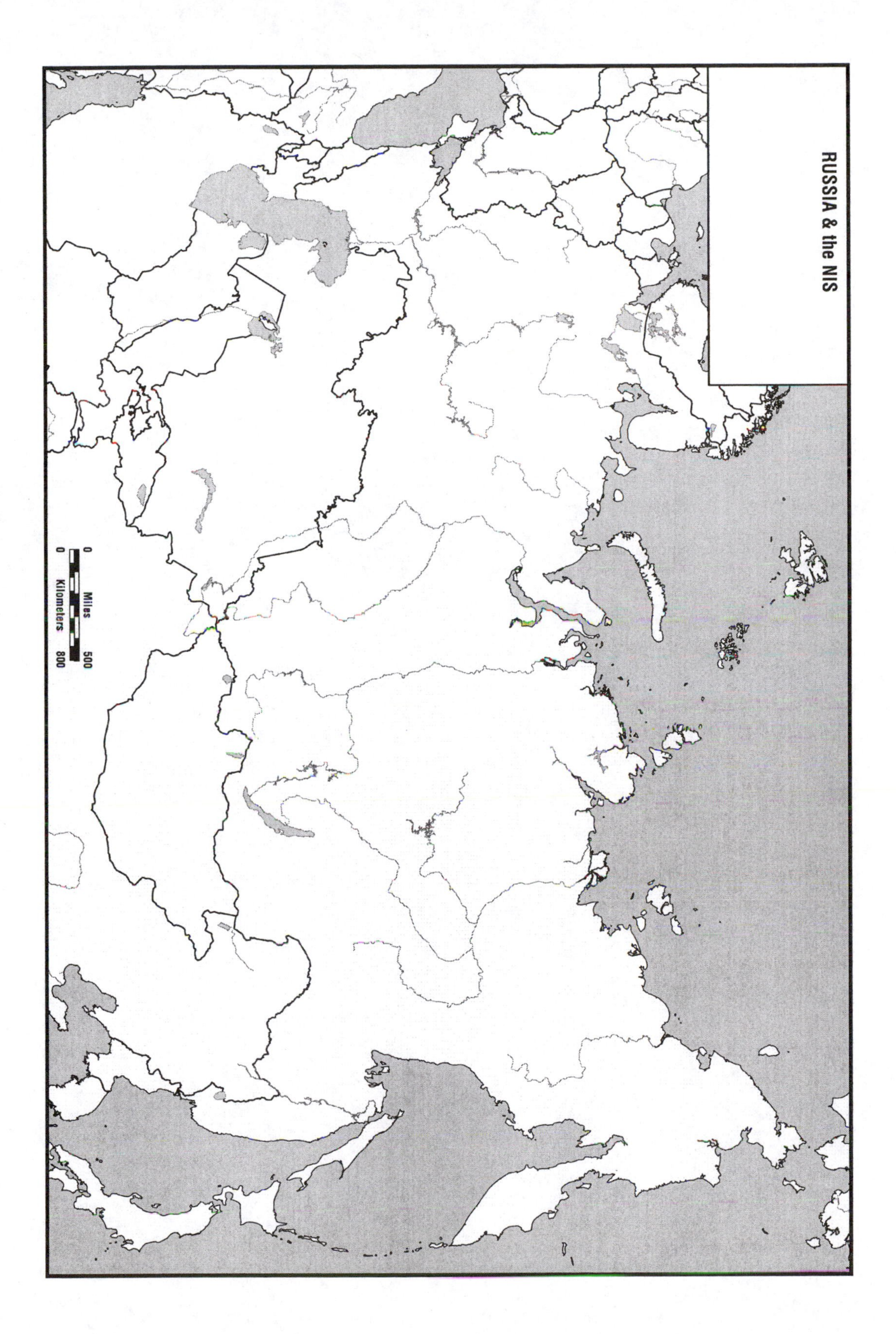
RUSSIA & the NIS
0 Miles 500
0 Kilometers 800

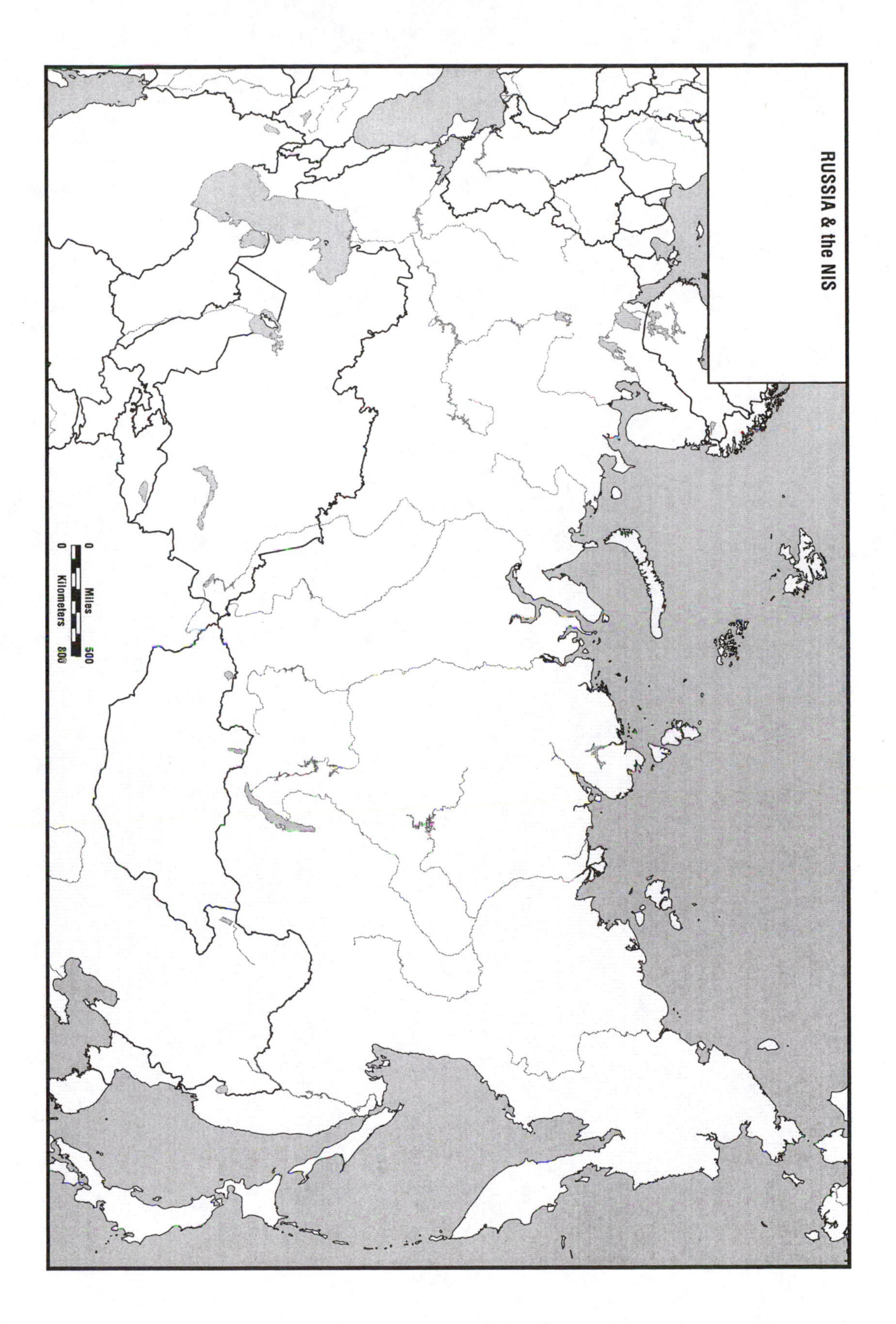
RUSSIA & the NIS
0 Miles 500
0 Kilometers 800

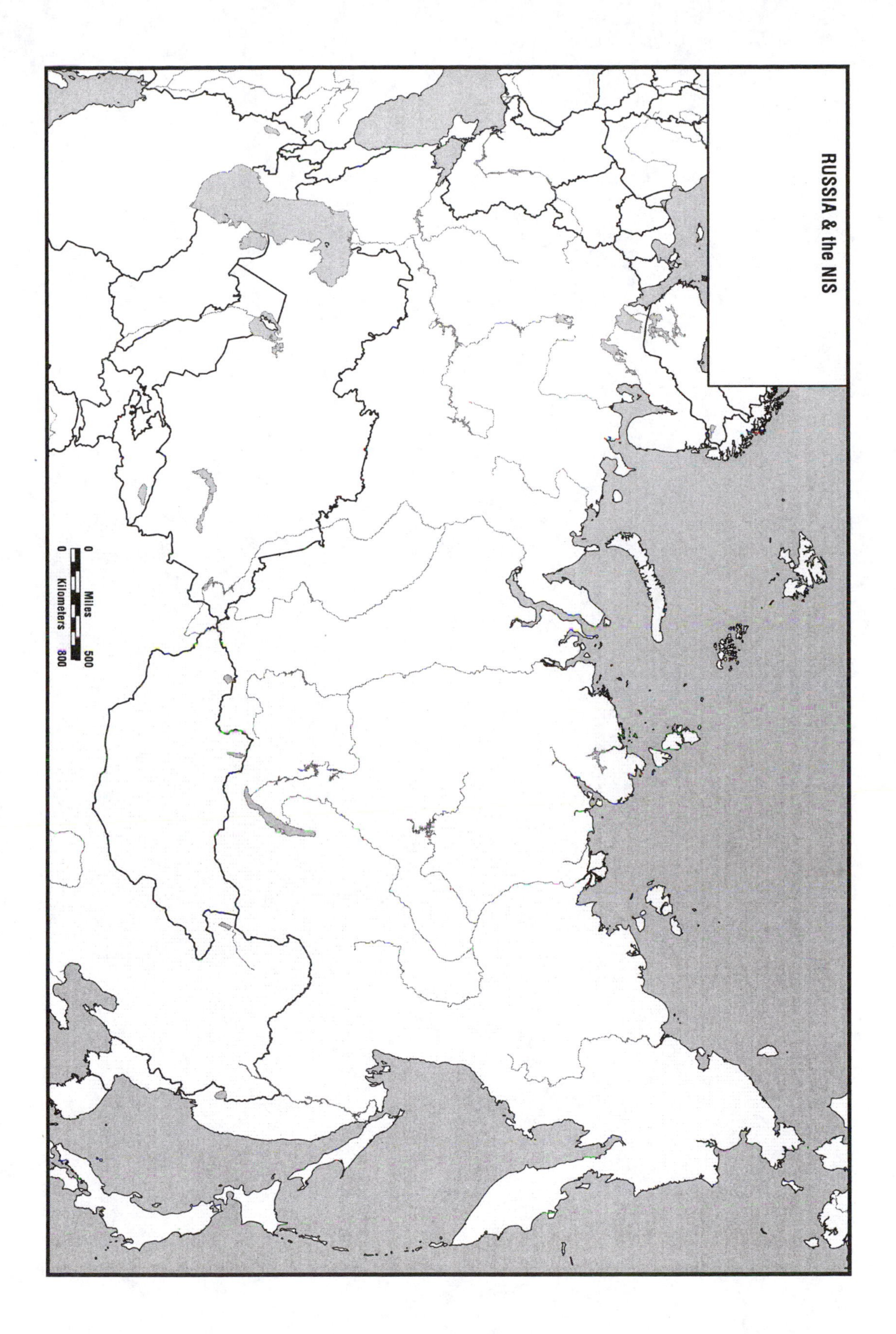
RUSSIA & the NIS
0 Miles 500
0 Kilometers 800

CHAPTER SIX
North Africa and Southwest Asia

LEARNING OBJECTIVES

After reading the chapter and working through this study guide, you should:

- Know how residents and governments are attempting to make this largely desert region inhabitable for a growing population. Understand the effects these attempts are having on people, the economy, and the environment; be able to speculate on how the growing population will affect water resources.
- Know how Islamic culture and beliefs affect government, law, and everyday behavior. Understand the causes and effects of the rise in Islamic fundamentalism.
- Know how early Western domination of this region affected economics, politics, and culture. Know some of the ways in which the effects of Western domination linger in this society today.
- Know some of the causes and effects of the restrictions on and seclusion of women. Understand why the patterns and degrees of seclusion vary greatly. Be able to appreciate how the changing role of women will affect culture, politics, and economics in this region.
- Know that oil revenue often results in significant income disparity. Know how countries' dependency on oil affects human well-being. Understand the challenges faced by countries in the region as they attempt to diversify their economies.
- Know the causes and effects of the various sources of hostility in this region: the Israeli-Palestinian conflict, tensions over the interpretation of Islam, access to water, and distribution of oil wealth.

KEY TERMS

The following terms are in **bold** in the textbook. In the space next to the term (or on a separate sheet or flash cards), you can fill in the definitions for reference or quiz yourself for exam review. Definitions are found in the glossary of the textbook.

Allah

cartel

Christianity

desertification

Diaspora

economic diversification

failed state

Fertile Crescent

hajj

intifada

Islam

Islamic fundamentalism (Islamism)

Islamists

Judaism

madrasa

monotheistic

Muslims

OPEC (Organization of Petroleum Exporting Countries)

pogroms

polygyny

protectorate

Qur'an (or Koran)

recharge area

salinization

seclusion

secular states

shari'a

sheikhs

Shi'ite (or Shi'a)

Sunni

terrorism

theocratic states

veiling

Zionists

REVIEW QUESTIONS

The following review questions are related directly to the textbook material. These questions can be used to help you prepare for an exam or you may want to read through the questions before you begin reading the textbook, quizzing yourself after you complete each section.

The Geographic Setting

1. Describe the climate-related function of the Atlas Mountains. How are the Atlas Mountains and Sahara Desert related? How do climate and human existence differ from one side of the mountains to the other?
2. What are some of the changes in prehistorical agricultural societies that began to transform the roles of men and women?
3. Although Mecca is maintained as the holy city of Islam, in keeping with the tradition of *hajj*, Islam has spread through much of the world. How did this diffusion occur? What were some of the routes and means of diffusion? What are some of Islam's contributions?
4. How have colonization, Western domination, and oil wealth affected this region's economy, politics, and environment?
5. The size, aridity, and population density of this region greatly affect human settlement. Where do people live and why? What problems might an expanding population cause in cities?
6. As labor-intensive agriculture declines, the need for children also decreases. Why then do fertility rates remain high in this region? What would it take to lower them?
7. The global economy is affected by more than just the buying and selling of goods and services. This region is heavily involved in the global economy, as workers are

leaving some areas and migrating into others. Who is coming and who is leaving? What are the reasons for the emigration and immigration? What are the effects on the region's economy as well as on the global economy?

8. War and environmental disasters are causing millions to end up in refugee camps, which often become semipermanent communities. What are the positive and negative outcomes of these settlements?

Current Geographic Issues

9. Although the shari'a has many different interpretations, the Five Pillars of Islamic Practice are standard. Briefly explain each of them.
10. In many places in this region, how women occupy public and private spaces is strictly regulated. What are some of these restrictions, and how do these standards vary across the region and by social class? What is the logic behind these restrictions, and how are they enforced?
11. Who are the Islamists? Why are revolutionary movements, including Islamic fundamentalism, on the rise? What does Islamism promise that secular governments do not?
12. What are some of the many economic and political barriers to peace and prosperity in this region?
13. What was the purpose of establishing OPEC? Has it been a success? What are the positive and negative effects of it? What are its limitations and how could OPEC be more responsive to the social needs of the countries it represents? What will happen to OPEC countries as oil supplies decline worldwide?
14. Economic diversification can bring prosperity and stability. What are some of the reasons this region has had trouble developing economically?
15. Structural adjustment programs (SAPs) are imposed to help reduce debt and cut the government's role in the economy. What are some of the positive and negative results of SAPs in this region?
16. A number of the ongoing hostilities in this region have their roots in a long history of outside interference. How do these conflicts affect economic and political progress, as well as social change?
17. How do Islamic beliefs affect the environment in this region? Why are environmental issues likely to increase in importance?
18. The region's aridity has caused a number of environmental problems. How have humans historically coped with water scarcity? How is this changing? Why are water problems getting worse, and how are countries dealing with these problems?
19. Desertification is an ecological disaster caused partly by human activity. What are some of the causes and effects of desertification?
20. Why is GDP per capita a poor measure of quality of life in this region? How is it possible for a country to have a low GDP but a high GEM or HDI? Give an example of a country in this situation. How is it possible for a country to have a high GDP but a low GEM or HDI? Give an example. What are several reasons why Israel is an exception to much of the inequality in this region?

CRITICAL THINKING EXERCISES

The following are "what if"-type questions that illustrate concepts you are learning from the textbook. These questions ask you to apply the ideas and principles you learned from the textbook to new situations.

1. Adaptations to the physical environment

The daily climate of this region can be harsh and dry; however, humans have learned to adapt and survive. Because water has always been in short supply in this arid region, cultural attitudes toward water differ greatly from those in North America.

- Temperatures in the Sahara can vary from below freezing to well over 100 degrees. Heating and air-conditioning systems are not widely available in North Africa and Southwest Asia. What adaptations do people in North Africa and Southwest Asia make to survive in such harsh conditions?
- In Rub'al Khali, strong winds create dunes more than 2000 feet high. How do people protect themselves from these persistent winds and blowing sand?
- Only three major rivers flow in this region, and rainfall is infrequent and light from November to April. How do farmers and the nonfarming population deal with this water scarcity?
- Reflecting on the answers you have given, how would *you* cope with these harsh conditions if you did not have all the resources available to you (e.g., central heat and air, four solid walls around your house, and running water in it)?
- Finally, take a guess at how many gallons of water you use in a day. Remember brushing teeth, flushing toilets, showering, washing dishes, and just drinking a glass of water. After you have made an estimate, find out how much water you really use on a typical day by going to the Web site: http://ga.water.usgs.gov/edu/sq3.html. Did the amount you use surprise you? It is estimated that a person needs about 50 liters (13.2 gallons) of water per day for sanitation, bathing, cooking, as well as for survival. How does your daily water usage compare?

2. Religion in daily life

The Five Pillars of Islamic Practice focus on how to live daily life.

- Consider the Five Pillars of Islamic Practice. If you follow a belief system other than Islam, do you have similar elements in your belief system? List the five most important "pillars" of your religion or belief system. Compare them to the Five Pillars of Islamic Practice.
- Do you find that your five "pillars" guide you through daily life, or do you practice your belief system only during certain times or in certain spaces?

3. Public spaces and restrictions on women

In this region, some women are active in public life and government, while others lead secluded domestic lives with little opportunity for education.

- Men and boys usually interact in public spaces, while women usually inhabit secluded domestic spaces. In many countries in this region, women have to be accompanied by a male relative in public. How would this affect your daily life? How would this affect the education and career path you have chosen?

- Give some examples of material culture designed to seclude women from the public gaze. Do you see any similar examples in your society? How would this seclusion affect your daily life and interactions?

4. Islamism in a globalizing world

Many Muslims see modern, Westernized culture as undermining important values. In reaction, religious fundamentalism is on the rise in many parts of the world.

- List some positive and negative impacts of each of the following perceived consequences of modernization: liberalization of women's roles, family instability, consumerism, and widening gaps between the rich and poor.
- Do you think Islamism and a reaffirmation of traditional values will reduce the negative impacts of modernization? Do you think Islamism can offer solutions to the region's problems and that things will be better if people return to strict interpretations of religious-based morality? Justify your response.
- How can modernization be combined with Muslim beliefs and values?
- What should be done about the rights of non-Muslims living in these societies?

5. Migration and refugees

Many immigrants in the region flee persecution, and then return after they are liberated. For example, 700,000 Jews were allowed to leave the former Soviet Union and enter Israel in the 1990s. However, because of the creation of the state of Israel in 1948, as many as 2 million Palestinians were displaced from their homeland.

- Why do you think Israel is such an important place for Jews?
- Where did the Palestinians go after leaving what is now Israel? What might their lives have been like because of this forced migration? What might life have been like for those Palestinians who stayed?
- What are the long-term effects of this migration on the Palestinians and on Israel?
- What would this forced migration be like for you to experience?

IMPORTANT PLACES

The following places are featured in the chapter. Make sure you can locate all of them on a map. Blank outline maps can be found on the textbook's Web site: www.whfreeman.com/pulsipher4e. Also, to prepare for quizzes and exams, write a few important facts about each place in the space provided.

Physical Features

1. Arabian Peninsula

2. Arabian Sea

3. Atlantic Ocean

4. Atlas Mountains

5. Black Sea

6. Caspian Sea

7. Dead Sea

8. Euphrates River

9. Gulf of Aden

10. Jordan River

11. Mediterranean Sea

12. Nile River

13. Persian Gulf

14. Red Sea

15. Rub'al Khali Desert

16. Sahara Desert

17. Tigris River

18. Zagros Mountains

Regions/Countries/States/Provinces

19. Algeria

20. Bahrain

21. Cyprus

22. Egypt

23. Gaza Strip

24. Golan Heights

25. Iran

26. Iraq

27. Israel

28. Jordan

29. Kuwait

30. Lebanon

31. Libya

32. The Maghreb

33. Morocco

34. Oman

35. Occupied (Palestinian) Territories

36. Qatar

37. Saudi Arabia

38. Sinai Peninsula

39. Sudan

40. Syria

41. Tunisia

42. Turkey

43. United Arab Emirates

44. West Bank

45. Western Sahara

46. Yemen

Cities/Urban Areas

47. Abu Dhabi

48. Algiers

49. Al Manamah

50. Amman

51. Ankara

52. Baghdad

53. Beirut

54. Cairo

55. Damascus

56. Doha

57. Gaza

58. Jerusalem

59. Khartoum

60. Kirkuk

61. Kuwait

62. Makkah (Mecca)

63. Medina

64. Muscat

65. Nicosia

66. Rabat

67. Riyadh

68. San'a

69. Tehran

70. Tripoli

71. Tunis

MAPPING EXERCISES

The following are three mapping exercises to improve your knowledge of the location of places, underscore why they are important, and clarify how they relate to each other. Some questions will ask you to locate places, compare maps, or fill in data; others will test your understanding of *why* you were asked to map the features that you did. Use the blank outline maps at the end of the chapter to complete these exercises. Additional blank outline maps can be found on the textbook's Web site: www.whfreeman.com/pulsipher4e.

1. Urbanization, female literacy, and opportunities for women

Rural women are often less secluded because they have many tasks to perform outside the home. However, women tend to have more education and job opportunities in urban areas.

- From the map of urban population (Figure 6.18), shade countries from light to dark that have the following percent of urbanization: 20-39 percent; 40-59 percent; 60-79 percent; and 80 percent or more.
- Using the table of human well-being rankings (Table 6.2), use a graduated symbol to represent the female literacy rate for all the countries in the region. For example, use a small triangle for 0-50 percent, a medium triangle for 51-75 percent, and a large triangle for 76-100 percent
- You will also need to refer to the map of women who are wage-earning workers (Figure 6.15) and the map of restrictions placed on women (Figure 6.24).

Questions

a. What relationship would you expect between a country's urbanization and: 1) the number of women who are earning wages, and 2) the restrictions placed on women? Why would you expect these relationships? Where do you see them?
b. What relationship would you expect to see between female literacy rates and: 1) the percentage of women who are earning wages, and 2) the restrictions placed on women? Where do you see these relationships?
c. Choose two countries that stand out as contradictory to what you would expect. Explain why this might be the case.

2. Oil wealth and human well-being

Oil wealth in the region is not evenly distributed. Many live in poverty with a less than desirable quality of life, while a few are extremely wealthy.

- Using Figure 6.27 (map of economic issues), carefully shade (in pencil) the oil and gas-producing areas.
- Also using Figure 6.27, cross-hatch (///) the countries that are OPEC members.
- Using Table 6.2 (human well-being rankings), write (inside the country border) the GDP per capita and HDI for each country in the region.

Questions

a. List the countries that have oil and gas production. What is the general relationship between GDP per capita and oil and gas production? Are there any anomalies? If so, explain why.
b. For each oil and gas-producing country, compare its GDP per capita to its HDI. Overall, in countries where GDP per capita is high, does HDI also rank high (remember that a high HDI ranking appears in the table as a low number)? Is it what you expected? Why or why not?
c. Overall, how is oil *positively* affecting the region's people, the economy, and politics? How is it *negatively* affecting the region's people, the economy, and politics?

3. The importance of (clean) water

Most people in this region understand the importance, even the necessity, of living near sources of water; thus, it is vital that these water sources are clean. Refer to the population density map (Figure 6.14) and examine patterns of population density.

Questions

a. Examining the map, how would you describe the pattern of population distribution? Is it concentrated in specific locations (e.g., coastal or interior; near water or deserts)?
b. Given the current distribution, where do you think the rapidly increasing numbers of people will live?
c. Next, examine the maps of human impact (Figure 6.35). Based on where you predicted the growing population would live, how might this growth be problematic? How might it further increase water shortages and desertification?

SAMPLE EXAM QUESTIONS

The following are sample questions to help you review for an exam. Answers are found in the back of this workbook.

1. The textbook raises the idea that as societies based on irrigated agriculture became more complex which of the following occurred?
   A. Gender roles were institutionalized.
   B. Nutrition quality of diets improved.
   C. Urban populations declined.
   D. Wars over water resources erupted.

2. Which of the following statements does NOT accurately portray the Ottoman Empire?
   A. It endorsed religious intolerance.
   B. Its capital city was Istanbul.
   C. Islam was the official religion.
   D. It was allied with Germany during World War I.

3. The textbook suggests that the recent decline in human fertility rates in urban areas in North Africa and Southwest Asia is largely attributable to which of the following?
   A. Women have been stripped of their right to an education.
   B. The shortage of basic necessities (food, health care) has worsened.
   C. Falling real incomes have convinced couples to have only one or two children.
   D. Women are choosing to work or study outside the home.

4. All *secular* countries exhibit which of the following characteristics?
   A. Its leader is a monarch who inherited the position.
   B. Its population is ethnically homogeneous.
   C. Its leaders are popularly elected.
   D. Its affairs are not directly influenced by religion.

5. Why are women in rural Islamic society less likely to be strictly secluded from public spaces?
   A. People living in rural places tend not to strictly subscribe to Islam and its restrictions on women's activities.
   B. Families in rural society are matriarchal, giving women the power to interpret Islamic strictures less rigidly.
   C. The tasks that women are obligated to carry out must be performed outside the home.
   D. As reflected in the large size of traditional rural families, taboos on female sexuality are less strict in rural places.

6. Which of the following is an accurate description of an *Islamist* movement?
   A. One that seeks to remove restrictions on Muslim women
   B. One that aims to end the overdependence on oil resources
   C. One that seeks to liberalize interpretations of the Qur'an
   D. One that favors the establishment of a state based on Islam

7. What is the main reason that the countries that experienced huge gains in oil wealth during the 1970s were not able to generate an industrial revolution in their domestic economies?
   A. They universally refused to purchase Western technologies and goods.
   B. They failed to reinvest their wealth abroad in more stable economies.

C. They did not sufficiently invest in basic human resources such as education and health care for people.
D. Oil wealth benefited mainly foreign "guest workers" who did not invest locally.

8. According to the textbook, which of the following has motivated countries in Southwest Asia and North Africa to try to attain self-sufficiency in food production despite the high costs of such development?
   A. United Nations-led trade embargoes
   B. Regional geopolitical tensions
   C. Dissolution of OPEC
   D. Rising levels of salinization

9. Even though structural adjustment programs (SAPs) have had a severely negative impact on low-income and poor people, why have many states in North Africa and Southwest Asia introduced such policies?
   A. Many governments in the region have curtailed their intervention in the economy to such a degree that most industries are no longer economically efficient.
   B. Social spending (education, heath care) in nearly all states has decreased, which has caused a sharp decrease in many people's real incomes.
   C. Domestic industries have fared poorly on world markets because subsidies are low and tariffs are too high.
   D. Most states bear crippling debt burdens that were acquired from loans taken to industrialize and modernize their societies.

10. In North Africa and Southwest Asia, which of the following effects is a direct result of the growth of industrial (modernized, mechanized) agriculture and manufacturing?
   A. Higher rates of human fertility
   B. Less dependence on domestic oil production
   C. Greater consumption of water resources
   D. Urban-to-rural migration

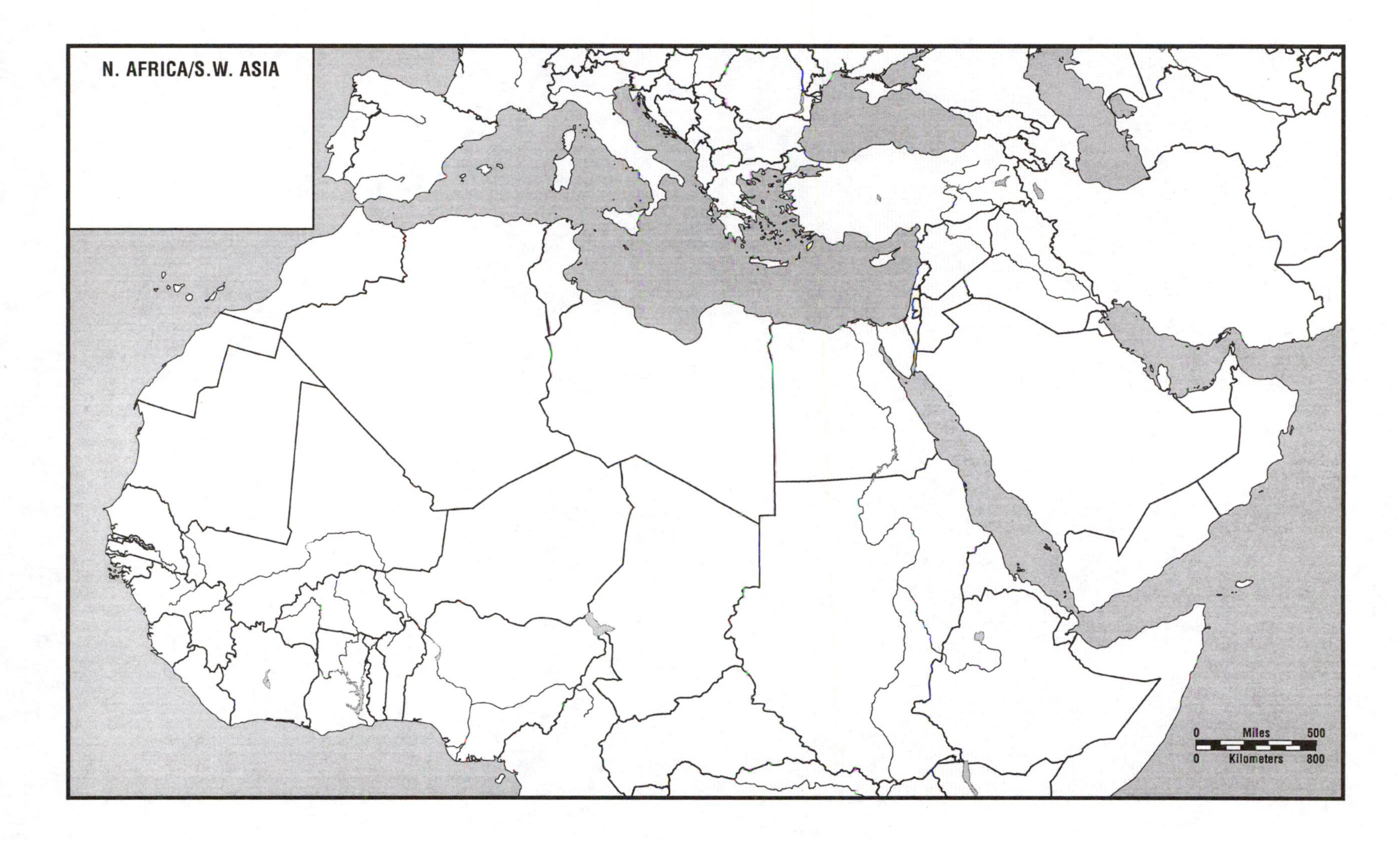
N. AFRICA/S.W. ASIA
0 Miles 500
0 Kilometers 800

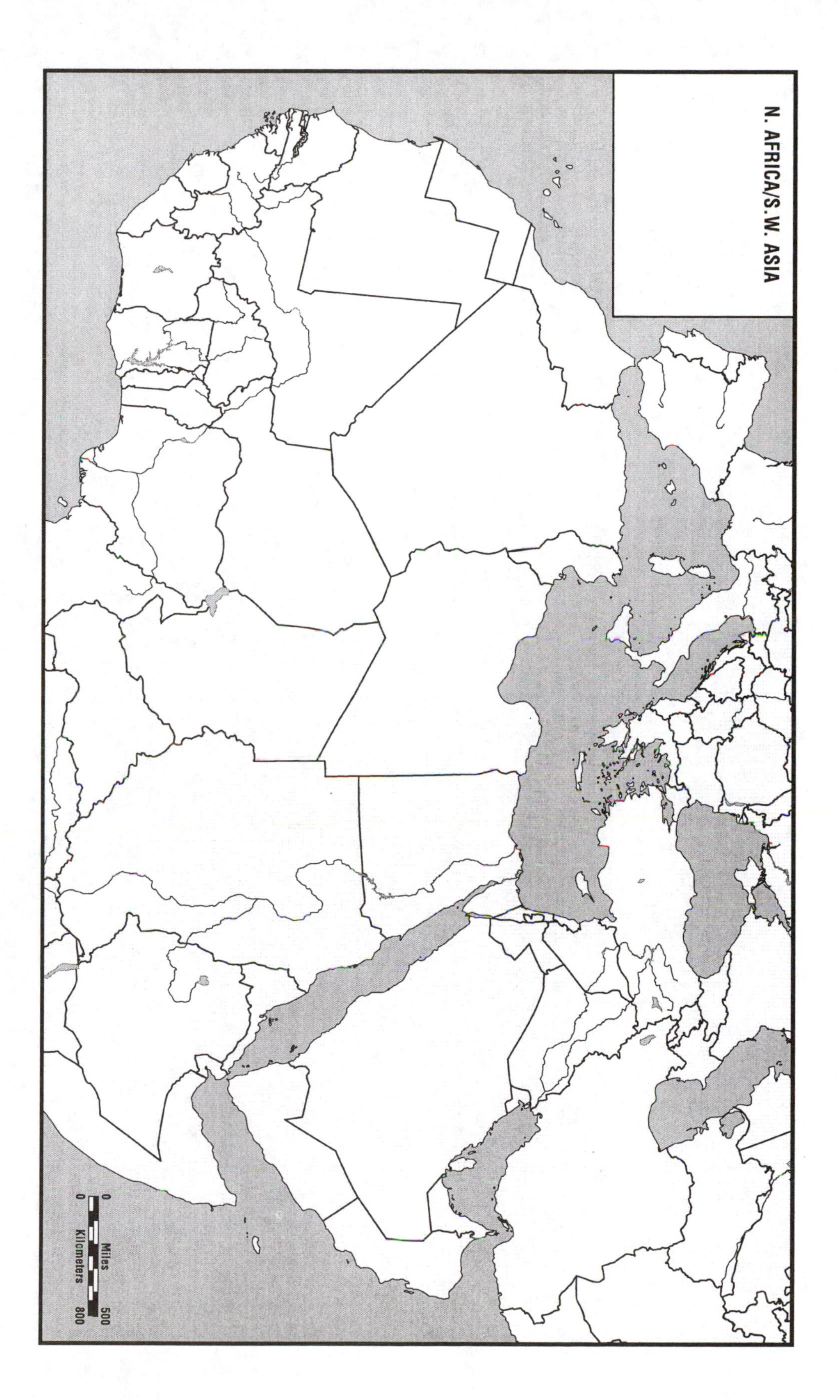
N. AFRICA/S.W. ASIA
Miles
0
500
Kilometers
0
800

CHAPTER SEVEN
Sub-Saharan Africa

LEARNING OBJECTIVES

After reading the chapter and working through this study guide, you should:

- Know how landforms and climate have challenged human beings and hindered Africa's development and connection to the outside world.
- Know how outsiders have influenced the region and understand the consequences of their actions. Understand why true independence from this domination is so difficult to achieve.
- Know why Africa's population is growing so rapidly. Know the consequences of such rapid growth in a region that is already suffering so many hardships. Understand what it would take for population growth to be reduced.
- Understand why diseases, including HIV-AIDS, are spreading so rapidly and are so difficult to control. Know the effects of these diseases on the population and on society.
- Know the reasons for the continued economic crisis in sub-Saharan Africa. Know what measures are being taken or considered to achieve economic sustainability. Know how political instability and ethnic strife compound these problems.
- Understand the importance of religion in daily life and how more recent religions (Islam and Christianity) were introduced, and the effects they are having. Know how colonization and the introduction of new religions have affected gender roles.
- Know the ways in which the leading environmental problems in this region - desertification, deforestation, diminishing wildlife, water scarcity, and pollution - are being addressed in the context of current economic and political challenges. Understand how humans have contributed to these problems.

KEY TERMS

The following terms are in **bold** in the textbook. In the space next to the term (or on a separate sheet or flash cards), you can fill in the definitions for reference or quiz yourself for exam review. Definitions are found in the glossary of the textbook.

African Union

agroforestry

animism

apartheid

carrying capacity

Common Market for Eastern and Southern Africa (COMESA)

currency devaluation

desertification

devolution

dry forests

East African Community (EAC)

Economic Community of Central African States (CEEAC)

Economic Community of West African States (ECOWAS)

escarpment

female circumcision

fragile environment

fusion

grassroots economic development

Horn of Africa

intertropical convergence zone (ITCZ)

laterite

leaching

lingua franca

Mano River Union (MRU)

mixed agriculture

neocolonialism

pastoralism

polygyny

Sahel

self-reliant development

shifting cultivation

Southern African Development Community (SADC)

syncretism

West African Economic and Monetary Union (UEMOA)

REVIEW QUESTIONS

The following review questions are related directly to the textbook material. These questions can be used to help you prepare for an exam or you may want to read through the questions before you begin reading the textbook, quizzing yourself after you complete each section.

The Geographic Setting

1. Why are Africa's landforms so exceptionally uniform? What effect has this had on transportation and trade?
2. What are the general characteristics of Africa's climates? What effect does the ITCZ have on climate? What challenges do these climates bring to human beings?
3. What were some of the causes and negative outcomes of internal and external slavery in Africa?
4. Why did European powers establish colonies in this region? What were the effects of colonization on human well-being, agriculture, politics, population, and the economy?
5. Why were colonial boundaries drawn? What are some of the lingering problems with the boundaries?
6. What is apartheid and what was its purpose? What were the Boers' role?
7. What is carrying capacity and what limits it in Africa? How might the carrying capacity differ if Africa were a wealthier region?
8. How does the rapid population growth of this region affect standards of living? Why are Africans still having large families? What are some factors that might cause growth rates to decline?

9. How might the HIV-AIDS epidemic affect population growth and life expectancy? Why is HIV-AIDS more difficult to control in Africa than in other, wealthier regions? What are some of the factors contributing to its rapid spread? What are some of the social consequences of this problem?

Current Geographic Issues

10. Why are raw materials an unstable base for the economy, and why is it difficult to reduce this dependency?
11. Why is Africa enacting structural adjustment programs (SAPs)? What are some of the benefits and drawbacks of SAPs in this region?
12. With regular employment difficult to find, many people are becoming involved in the informal economy. What are the benefits and drawbacks?
13. What are the benefits and limitations of regional economic integration and grassroots rural economic development?
14. Why are politics so problematic in this region? Is this changing? Why is democracy so hard to achieve in Africa? What are some signs that democracy is on the rise?
15. Although most people live in villages, mass migration to cities is occurring. Why is this the case, and what is the effect of rapid urbanization on large cities and their infrastructure?
16. What are the common characteristics of indigenous belief systems? How did Islam and Christianity spread to this part of the world?
17. How do the terms ethnicity, race, color, and culture differ, yet relate to each other?
18. What are some possible causes and outcomes of desertification? How are humans compounding the natural cycles and shifts of vegetation and ecosystems? How might desertification be slowed or halted in this region?
19. Why is water scarce and unsafe in this region? What are some solutions to Africa's water problems?
20. Considering GDP, HDI, and GDI rankings, what is the status of human well-being in this region? What are some of the reasons for these rankings?

CRITICAL THINKING EXERCISES

The following are "what if"-type questions that illustrate concepts you are learning from the textbook. These questions ask you to apply the ideas and principles you learned from the textbook to new situations.

1. Ethnocentrism and biased views

It is impossible to generalize about Africa because environments, political systems, and ethnic groups are so diverse; thus, it is often the subject of misrepresentation and unwarranted, often negative, generalizations. The language used to describe Africa has been particularly prone to ethnocentrism.

- Before you start reading the textbook for this region, write down ten characteristics (generalizations) about it.
- After you've read the chapter and discussed it in class, look at your list again. Would you consider any of the terms you used to be ethnocentric? Why do you think this is the case? Have these views changed after reading the chapter?

2. Apartheid

Apartheid laws were enacted in 1948 to reinforce the long-standing segregation in South Africa. Everyone except whites had to carry passbooks and live in racially segregated areas or homelands. The fight to end racial discrimination in South Africa actually began before apartheid laws were even introduced. When apartheid finally ended only about a decade ago, one of the most prominent resisters of apartheid was elected president.

- Consider the historical and current relationships that Caucasians have with African Americans, Hispanic Americans, Asian American, and Native Americans in the United States. How are these situations similar to apartheid in South Africa?
- How have race relations changed in the United States since the official end of segregation in the 1960s? How have they changed in South Africa since the end of apartheid? Be sure to consider the fact that laws may have changed, but there is a lag time in how people's attitudes and actions change.
- Do you think South Africa should use the United States as a model for desegregation efforts? Why or why not?

3. Africa's importance in the global economy

Africa is often assumed to be unimportant to the global economy because it is poor and makes only a small contribution to the world's commerce. However, Africa is actually inextricably linked to the global economy because of its reliance on exports and imports.

- Do you think Africa plays a unique role in the global economy? Why or why not?
- Consider the objects in your room/apartment/house. What products do you think might be made from African raw materials?
- Using the textbook, Internet, or other sources, identify at least ten of Africa's main exports. Are you surprised? Why or why not? Start with the following Web sites:
  - www.cia.gov/cia/publications/factbook
  - reportweb.usitc.gov/africa/trade_data.jsp
- With this knowledge, reconsider what objects in your room/apartment/house could be from Africa. Keep in mind that you may not have many *finished* products from Africa, but many things you own might be made from Africa's raw materials. List at least ten things that are or could be made from African materials.
- After completing this exercise, reassess your first answer: Do you think Africa plays a unique role in the global economy? Why or why not?

4. Settlement patterns

Human settlements take many forms and the various types of living arrangements reflect how people relate to one another economically, politically, and socially. Although more than 65 percent of sub-Saharan Africans live in rural areas, the trend toward more urban settlements is growing.

- Make a list of the positive and negative aspects of living in rural and urban areas in your country. Consider economic, political, and social characteristics. What are some of the differences between living in rural and urban areas in your country compared to rural and urban areas in a sub-Saharan African country?
- Would you choose to live in an urban or rural setting in your country? How about in sub-Saharan Africa? Why? Are your reasons because of push factors or pull factors?
- If you live in an urban setting, what are some of the ways your life would be different if you lived in a rural setting? Or, if you live in a rural setting, how would your life be different if you lived in an urban setting?

5. Female genital mutilation

Multiple and complex symbolic meanings may explain this deeply ingrained custom. Although many see it as an extreme human rights abuse, it has significant importance to others.

- Why is female genital mutilation (FGM) so important and ingrained in certain value systems? What are some of the arguments defenders of this practice make to support it?
- What are some of the negative consequences?
- Would this custom be acceptable in the United States? Why or why not? Do you think it is practiced in the United States? Using the Internet or other resources, find out if it is practiced in the United States.
- Does the United States have any similar rituals?
- Should the practice of FGM end? Defend your response. If you answered yes, what can be done to eliminate FGM?

IMPORTANT PLACES

The following places are featured in the chapter. Make sure you can locate all of them on a map. Blank outline maps can be found on the textbook's Web site: www.whfreeman.com/pulsipher4e. Also, to prepare for quizzes and exams, write a few important facts about each place in the space provided.

Physical Features

1. Atlantic Ocean

2. Cape of Good Hope

3. Congo Basin

4. Congo (Zaire) River

5. Ethiopian Highlands

6. Great Rift Valley

7. Horn of Africa

8. Indian Ocean

9. Kalahari Desert

10. Lake Chad

11. Lake Malawi

12. Lake Tanganyika

13. Lake Victoria

14. Mount Kenya

15. Mount Kilimanjaro

16. Namib Desert

17. Niger River

18. Orange River

19. Red Sea

20. Sahara Desert

21. Sahel

22. Zambezi River

Regions/Countries/States/Provinces

23. Angola

24. Benin

25. Botswana
26. Burkina Faso
27. Burundi
28. Cameroon
29. Cape Verde Islands
30. Central African Republic
31. Chad
32. Comoros
33. Congo (Brazzaville)
34. Congo (Kinshasa)
35. Côte d'Ivoire
36. Djibouti
37. Equatorial Guinea
38. Eritrea
39. Ethiopia
40. Gabon
41. Gambia
42. Ghana
43. Guinea
44. Guinea-Bissau
45. Kenya
46. Lesotho
47. Liberia

48. Madagascar

49. Malawi

50. Mali

51. Mauritania

52. Mauritius

53. Mozambique

54. Namibia

55. Niger

56. Nigeria

57. Réunion

58. Rwanda

59. São Tomé and Principe

60. Senegal

61. Seychelles

62. Sierra Leone

63. Somalia

64. South Africa

65. Swaziland

66. Tanzania

67. Togo

68. Uganda

69. Zambia

70. Zimbabwe

Cities/Urban Areas

71. Abidjan

72. Abuja

73. Accra

74. Addis Ababa

75. Antananarivo

76. Asmera

77. Bamako

78. Bangui

79. Banjul

80. Bissau

81. Brazzaville

82. Bujumbura

83. Cape Town

84. Conakry

85. Dakar

86. Dar es Salaam

87. Djibouti

88. Freetown

89. Gaborone

90. Harare

91. Johannesburg

92. Kampala

93. Kigali

94. Kinshasa

95. Lagos

96. Libreville

97. Lilongwe

98. Lomé

99. Luanda

100. Lusaka

101. Malabo

102. Maputo

103. Maseru

104. Mbabane

105. Mogadishu

106. Mombasa

107. Monrovia

108. Moroni

109. Nairobi

110. N'Djamena

111. Niamey

112. Nouakchott

113. Ouagadougou

114. Port Louis

115. Porto-Novo

116. Praia

117. Pretoria

118. São Tomé

119. Victoria

120. Windhoek

121. Yamoussoukro

122. Yaounde

MAPPING EXERCISES

The following are three mapping exercises to improve your knowledge of the location of places, underscore why they are important, and clarify how they relate to each other. Some questions will ask you to locate places, compare maps, or fill in data; others will test your understanding of *why* you were asked to map the features that you did. Use the blank outline maps at the end of the chapter to complete these exercises. Additional blank outline maps can be found on the textbook's Web site: www.whfreeman.com/pulsipher4e.

1. Population growth, education, and HIV-AIDS

HIV-AIDS is the most severe public health problem in sub-Saharan Africa. It is having dramatic effects on population growth and life expectancy patterns. However, education is beginning to play a changing role in the spread and control of HIV-AIDS.

- Shade those countries that have 5 percent or more of adults with HIV-AIDS (Figure 7.17). Use a lighter shade for countries with 5-15 percent and a darker shade for countries with 15-34 percent. Label these countries.
- From the current year's "World Population Data Sheet," found at www.prb.org, look up the countries you labeled. Draw a hatch pattern over the countries that have over 2 percent rate of natural increase.
- Finally, use a graduated symbol (e.g., small to large circles), and draw a symbol in each of these countries to illustrate its female literacy (Table 7.1).

Questions

a. What is the current relationship between HIV-AIDS and population rate of natural increase? Explain why this is the case.
b. If the situation stays the same, how will HIV-AIDS affect life expectancy and population growth of this region 20 years from now? How could this situation be changed?

c. Which of these countries do you think will experience the greatest *decrease* in population growth rate? Why?

d. Examining the literacy data, how might increased education affect countries that have high HIV-AIDS rates? How will it affect those with high growth rates? Explain why. Some countries with high HIV-AIDS rates *do* have high rates of literacy; why do you think they still have high rates of HIV-AIDS?

2. Democracy and the provision of basic needs

In examining the daily suffering of many Africans, many wonder if African governments should meet their citizens' basic needs for food, shelter, and health care *before* they open up to a democratic form of government.

- Using Figure 7.35, shade the countries that had democratically elected governments in 1970. Choose a different color and shade the additional countries that had democratically elected governments in 2006. Cross hatch all countries that had democratically elected governments in 1970, but *not* in 2006. Label all the countries you shaded.
- Use graduated symbols to illustrate the Human Development Index as either "low," "medium," or "high" for all countries (Table 7.1).

Questions

a. Based on your map, which meets basic needs better, democratic or undemocratic governments? Why do you think this is the case? What can you interpret about the connection between HDI and democracy for those who had democratic elections in 1970 but *not* in 2006?

b. In light of the suffering of Africa's people, explain whether you think African governments should meet basic needs like food, shelter, and health care before they focus on achieving democracy.

c. What are some of the difficulties governments face in providing for citizens?
   - How can countries provide for needs when they have a very limited tax base from which to draw?
   - Many of these countries were established as democracies at independence, yet authoritarian presidents have caused major problems. How do you make a corrupt dictator take care of the basic needs of the people?

d. Explain if you think Africa would be better off left alone, or should other countries help provide basic needs for Africans?

3. The limits on carrying capacity

Carrying capacity depends on physical factors, including water supply and quality, soil condition, and disease, as well as cultural, social, economic, and political factors, including agriculture, wealth, and political unrest.

- Shade (in yellow) the countries with 5 percent or more of their population infected with HIV (Figure 7.17).
- Choose a symbol to draw in each country with GDP per capita less than $1000 (Table 7.1).

- Using the regional map of sub-Saharan Africa (Figure 7.1), outline (in light brown) and label the Kalahari, Namib, and Sahara deserts.
- Using Figure 7.1, trace and label the major rivers in the region with a heavy blue line: Niger, Orange, Congo (Zaire), Nile, White Nile, Blue Nile, and Zambezi.
- Using the map of human impact (Figure 7.45), shade areas that have high impact and medium-high impact on the land.

Questions

a. Make a list of at least five countries on your map that appear to have the potential for high carrying capacity (i.e., low human impact on the land, not in a desert, low percent of population with HIV-AIDS, and a relatively high GDP per capita).
   - From the current year's "World Population Data Sheet," found at www.prb.org, add their rate of natural increase to your list.
   - Is rapid population growth occurring? What implications might this population change have on these countries?

b. Next, from the "World Population Data Sheet," make a list of the ten fastest growing countries in sub-Saharan Africa. Label these ten countries on your map.
   - Examining your map, does the carrying capacity of each of these ten rapidly growing countries appear to be high or low? Write high or low next to each.
   - What implications might this rapid growth have on the countries that already have a low carrying capacity?

SAMPLE EXAM QUESTIONS

The following are sample questions to help you review for an exam. Answers are found in the back of this workbook.

1. Which of the following is the most accurate general description of the geomorphology of the African continent?
   A. It resembles a dinner plate with a sunken interior ringed by weathered mountains.
   B. It is comparable to a rumpled carpet with high mountains and narrow valleys.
   C. It appears as an inclined plane, sloping from north to south.
   D. It resembles a raised platform edged by narrow coastlines.

2. The textbook makes the point that Africa's climate creates many challenges for humans who live there. Which one of the following is NOT mentioned as a challenge?
   A. Soil is not as fertile in drier tropical climates.
   B. Excessive rain leads to depression and higher suicide rates.
   C. Diseases are easily cultivated and spread in the warm and wet climate.
   D. In moist and hot climates the soil decays rapidly.

3. Which of the following does NOT reflect the lasting influence of the colonial period on African societies?
   A. Many independent African governments are undemocratic.
   B. Europeanized elites usually dominate government and the economy.
   C. African economies depend on the export of raw materials.
   D. African governments effectively redistribute wealth to an impoverished majority.

4. Which of the following reflects the meaning of the term "carrying capacity"?
   A. Maximum number of people a given territory can support sustainably
   B. Minimum size an economy must be to support a population of a given size
   C. Minimum amount of food a household must possess to survive for one season
   D. Maximum amount of land a society can cultivate to support both itself and trade

5. Which of the following does NOT explain why birth rates in sub-Saharan Africa remain high?
   A. Children are viewed as a link between the past and future.
   B. Children can provide labor on family farms.
   C. High infant death rates encourage couples to have more children.
   D. Adequate vaccinations for most diseases are now widely available.

6. The economic crisis in sub-Saharan Africa is a product of which of the following?
   A. Debt has accumulated from years of low economic growth.
   B. The prices of imported manufactured goods have rapidly decreased.
   C. European countries have refused to pay off their debts to the region.
   D. Most of the resources (for export) of the region have been depleted.

7. Which of the following reflects a problem associated with the informal economy in sub-Saharan Africa?
   A. Profits are unreliable.
   B. It produces more tax revenue for the state.
   C. Profits are stable and have led to increased living standards.
   D. Incomes of those it employs are increasing.

8. Among the following, which is the most obvious European legacy at the root of many armed conflicts in sub-Saharan Africa?
   A. National borders
   B. Ethnic discrimination
   C. Environmental destruction
   D. Socialist values

9. According to the long-standing division of labor in sub-Saharan Africa, which one of the following tasks is typically NOT the responsibility of women?
   A. Caring for children and elderly
   B. Obtaining and transporting water for the house
   C. Gathering and carrying firewood for household use
   D. Clearing and turning the soil for cultivation

10. Which of the following is NOT among the general issues concerning water in sub-Saharan Africa?
    A. Water must almost always be carried from a source external to the home.
    B. Women procure virtually all water consumed by households.
    C. Plumbing and sewage treatment are usually not available.
    D. Development has increased per capita supplies of water.

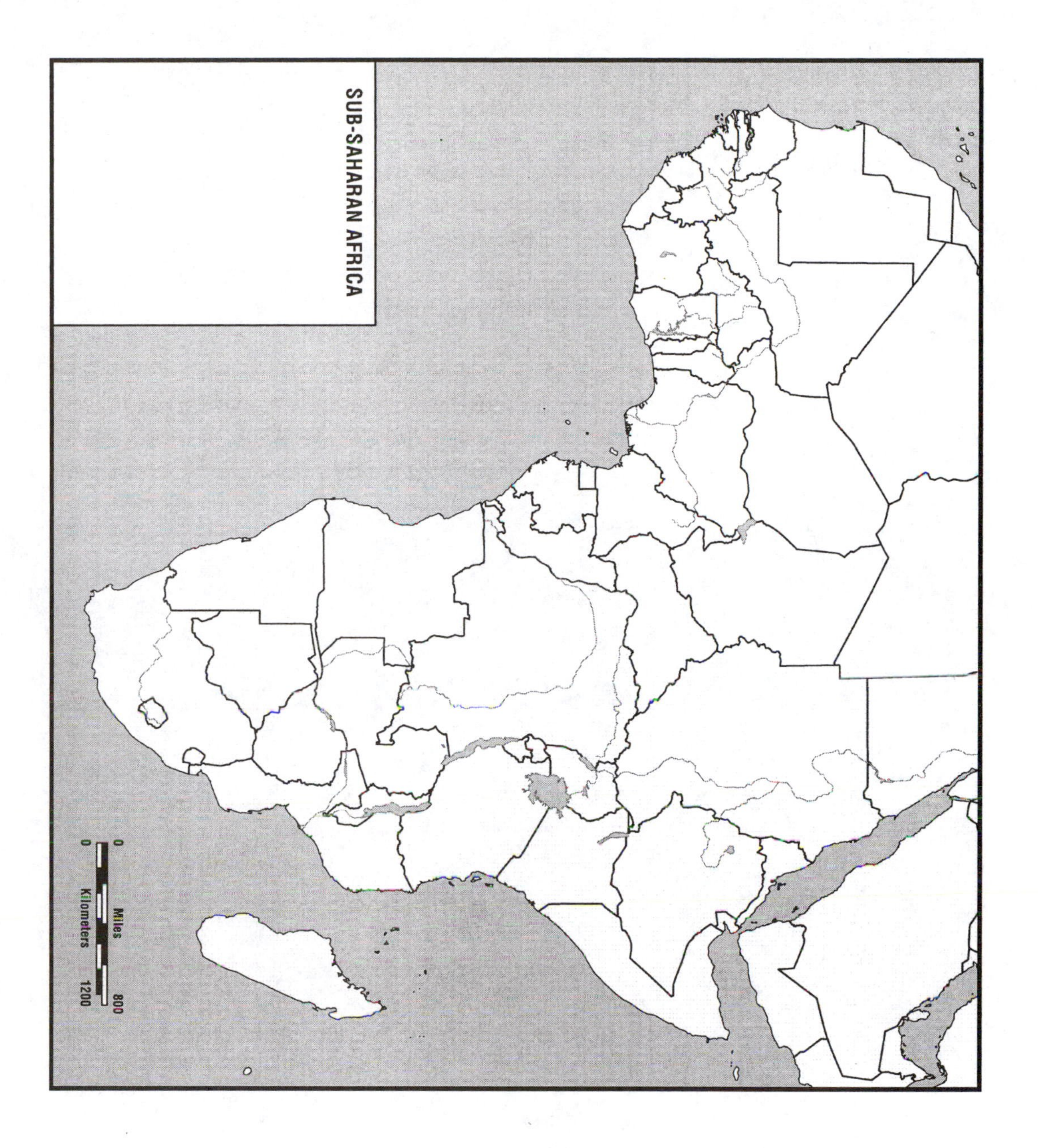
SUB-SAHARAN AFRICA
0 Miles 800
0 Kilometers 1200

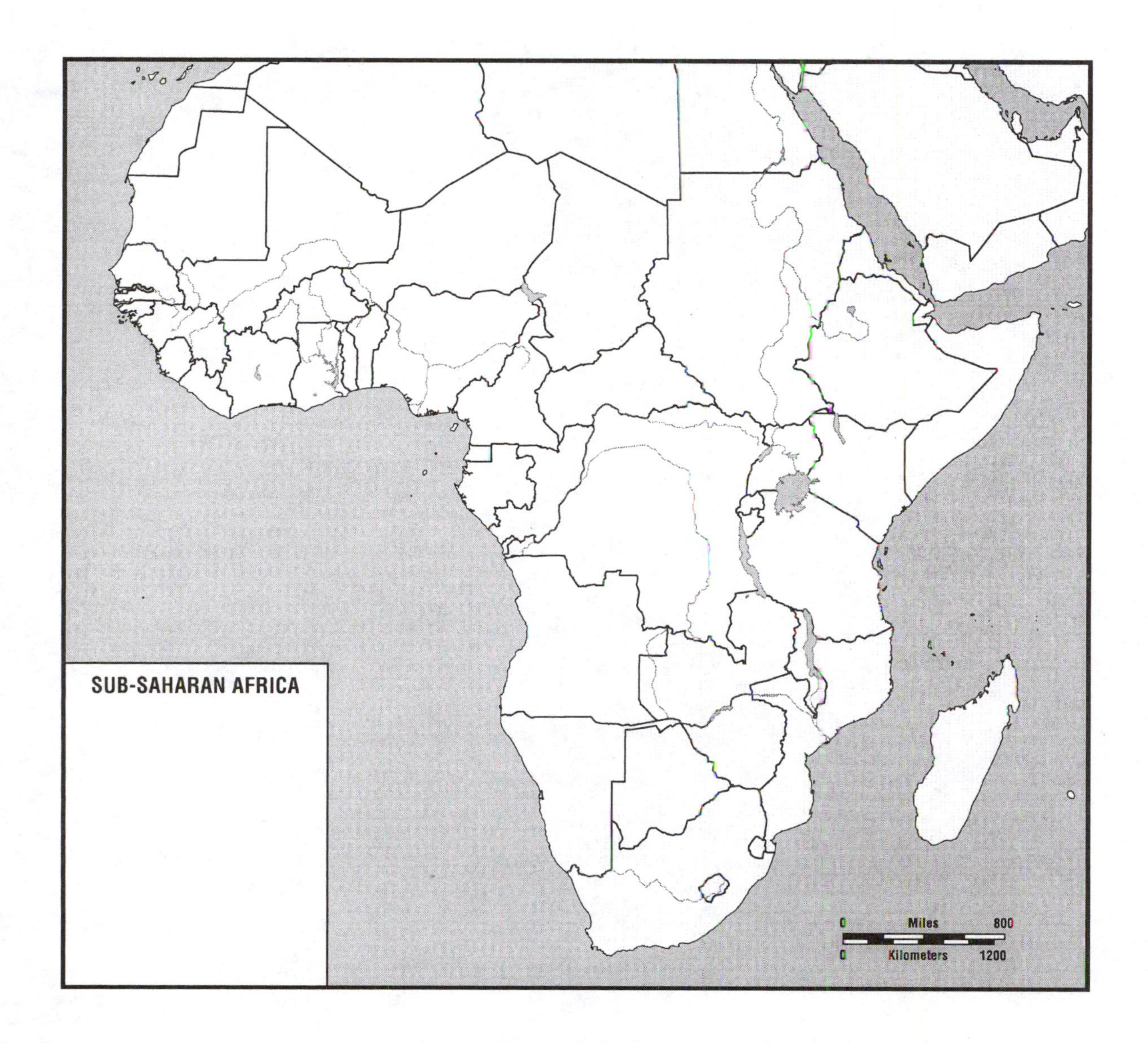
SUB-SAHARAN AFRICA
0 Miles 800
0 Kilometers 1200

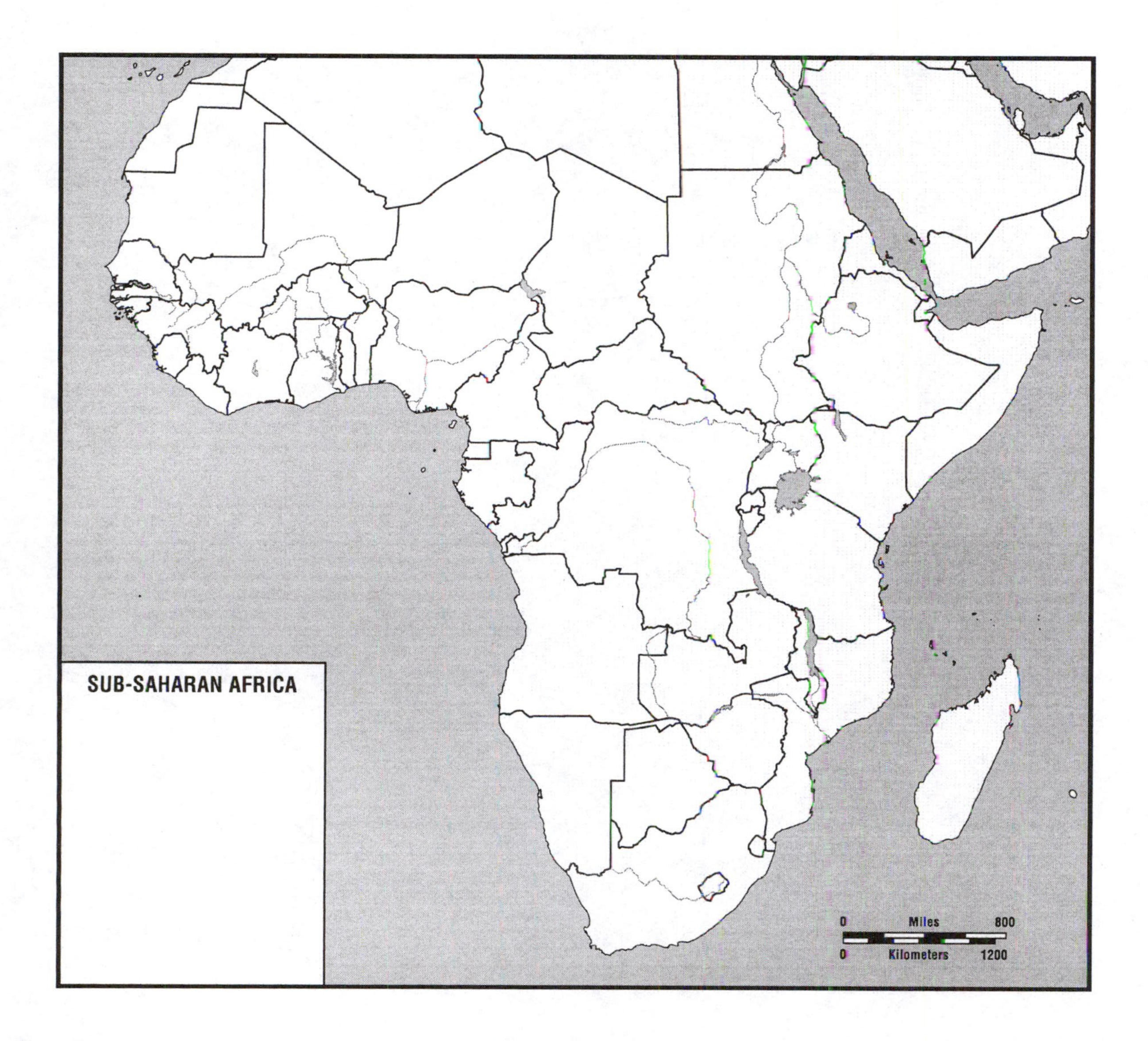
SUB-SAHARAN AFRICA
0
Miles
800
0
Kilometers
1200

CHAPTER EIGHT
South Asia

LEARNING OBJECTIVES

After reading the chapter and working through this study guide, you should:

- Know how climate (especially monsoons) affects agriculture, the economy, health, and daily life.
- Be able to identify the residual positive and negative influences of British colonization and subsequent independence.
- Know why population is growing rapidly among the very poor majority. Understand the potential consequences of such rapid growth. Know what can be done to curtail this growth and increase quality of life.
- Understand the historical importance of the caste system and how its role has changed over time.
- Know the overall status of women and its effect on quality of life. Know why purdah is desired by some and not by others. Know how freeing women from purdah will affect them, their families, and their communities.
- Understand why there are such disparities of wealth in the region. Know what is being done to counteract these inequalities.
- Know why, although often praised as a bastion of democracy, South Asia still has many tensions and potential for conflict.
- Know some of the environmental issues facing South Asia and why they are so difficult to resolve.

KEY TERMS

The following terms are in **bold** in the textbook. In the space next to the term (or on a separate sheet or flash cards), you can fill in the definitions for reference or quiz yourself for exam review. Definitions are found in the glossary of the textbook.

Adivasis

agroecology

Aryans

Brahmins

bride price

Buddhism

caste

Chipko movement

civil disobedience

communal conflict

dowry

green revolution

Harappa culture

Harijans

hearth

Hindi

Hinduism

Indian diaspora

Indus Valley civilization

Jainism

jati

Kshatriyas

microcredit

monsoon

Mughals

Parsis

Partition

purdah

regional conflict

religious nationalism

Sikhism

social forestry movement

subcontinent

Sudras

Taliban

Vaishyas

varna

REVIEW QUESTIONS

The following review questions are related directly to the textbook material. These questions can be used to help you prepare for an exam or you may want to read through the questions before you begin reading the textbook, quizzing yourself after you complete each section.

The Geographic Setting

1. Describe the resulting landscape from the collision of tectonic plates on this region.
2. Describe the role the intertropical convergence zone (ITCZ) and monsoons play in the climate of this region. What are the consequences of such severe weather on the landscape and people?
3. What is the Indian diaspora? How does it enhance global connections? How does it affect South Asia?
4. Trace the path of colonization in this region. What are some of the positive and negative legacies of the Aryans, Mughals, and British?
5. Why was India partitioned in 1947 and what were some of the outcomes? What were the results of the civil war in Pakistan in 1971? What are relations between India and Pakistan like today?
6. What are some of the negative effects of the high densities and growing population on both rural and urban areas? How is the region's growing population affecting quality of life? Why does population continue to grow in this region, and what will it take to slow this growth?

Current Geographic Issues

7. What are some of the roots of the great degree of ethnic and linguistic diversity?
8. Why is the relationship between Hindus and Muslims so complex? How do relationships differ in rural and urban areas?
9. What is the caste system? What activities does it affect? What are the reasons for its endurance in some areas and decline in others? What are some of the positive and negative effects it has had on societies?
10. Explain the general status of women in South Asia. Who does and does not observe purdah? How and why is this changing?
11. What is the difference between a bride price and a dowry? How are "bride burnings" or "dowry killings" and female infanticide related to dowries?
12. What is the cause of the startling economic contrasts in this region? What are the effects?
13. Give at least five general characteristics of agriculture in this region. Why was the green revolution seen as a necessity? What were its positive and negative outcomes?
14. What sector(s) of the economy is/are contributing to economic globalization in this region? What are the reasons for both the fear of and desire for globalization in this region?
15. Explain microcredit. How does it work and what are the benefits to village and rural women in particular?
16. Religious nationalism and regional conflicts are detrimental to the stability of the region. Why are they currently on the rise?
17. What are some signs that democracy is on the rise in this region?
18. Why has deforestation become an issue for both urban and rural areas? Who is losing and who is benefiting from the deforestation?
19. What are some of the controversial water issues in this region? Why are they so difficult to solve? How are the poor, in particular, affected by these problems?
20. How do South Asians survive with such low GDPs? Why are HDI rankings so low?

CRITICAL THINKING EXERCISES

The following are "what if"-type questions that illustrate concepts you are learning from the textbook. These questions ask you to apply the ideas and principles you learned from the textbook to new situations.

1. Coping with monsoon rains

The seasonality of monsoons and the amount of rainfall from them dictate the lives of many people in South Asia.

- From the Web site www.worldclimate.com, find the seasonality of rains in your area and the rainfall it receives (choose the nearest city if your city does not have data).
- How does the *seasonality* in your area compare with monsoon seasons in South Asia? What if you experienced the drastic seasonality experienced in parts of South Asia? How would you cope (think about agriculture, transportation, economy, daily life, etc.)?

- How does the *rainfall amount* in your area compare with the amount in places in South Asia? Collect data for several cities in South Asia. Consider how this much precipitation would affect you and the area in which you live.
  - Would you have to make changes in your lifestyle? What kind of changes?
  - Are we in the industrialized world (with solid structures and early warning systems) so removed from climatic anomalies that rainfall amount wouldn't really matter? Justify your response.

2. Are there other forms of the caste system?

In South Asia, the caste system divides society into a social hierarchy. Traditionally, a person is born into a caste, which cannot be changed. It affects nearly all aspects of daily life.

- All human groups have deeply ingrained concepts of relative social status. Consider the caste system in South Asia as well as social differentiation in your society.
  - Does the society you live in have anything that resembles the caste system (think about race, gender, age, occupation, location, etc.)? Make a list of categories in one of these "caste systems" in your society.
  - How is "caste" indicated (think about clothes, hairstyle, body decorations, manner of speaking, material possessions, space occupied, gender, religion, etc.)?
  - What "caste" would you fit into in the categories you listed?
  - What effects does this system have on your everyday life and society in general?
- Many South Asians converted from Hindu to Islam to escape life as a member of a low-status caste. Consider if you would be willing to change your belief system in order to escape the "caste" you are in.

3. Purdah and the status of women

Purdah is the practice of concealing women from the eyes of nonfamily men. This is a widely accepted practice in many parts of the world and has dramatic effects on women's lives and relationships.

- Why is purdah so accepted in this region? Do you think it would be acceptable in your society? Why or why not? What can you think of in your society that compares to the practice of purdah?
- How would living under purdah affect your life and the choices you make?
- Some women choose seclusion in South Asia. What reasons do they give for this choice? Under what circumstances might *you* want to practice purdah?
- What are the benefits of *ending* the practice of purdah to men, women, and children?

4. Microcredit at your fingertips

South Asians have developed innovative ways to help the poor. In 2006, Dr. Mohammed Yunnis was awarded the Nobel Peace Prize for his creation the Grameen Bank in the 1970s. Not only has the Grameen Bank lent nearly $6 billion to over 6 million borrowers, many other organizations, both international and local, have found success with microlending.

- Research microlending/microcredit in your local community (or the nearest large city). Were you surprised at what you found? Does a local microcredit group or national organization serve your community? What types of projects are being funded and how is repayment organized?
- Newer international microcredit organizations are also finding success. For example, Kiva, started by a young married couple in San Francisco in 2004, allows people all over the world to lend in increments of $25 through PayPal. Internet technology allows you to see photos of borrowers and read their stories before selecting a loan to fund. While your borrower is working hard to get out of poverty in Tanzania, Paraguay, Azerbaijan, or elsewhere, you are continually updated as they repay their loan to your PayPal account. Research Kiva (www.kiva.org) and similar international lending institutions (start with www.slate.com/id/2161797). What types of projects are being funded and how is repayment organized?
- Do you think this is a promising strategy for helping the working poor? Does it appear to be a success? Justify your answers.

5. Environmental impacts as a tourist

Deforestation is not new to the region of South Asia, but it is becoming worse as tourism and commercial uses of forest products increase.

- Think about the last trip you took to a natural area, perhaps to go hiking or camping. How were natural resources sacrificed (if even in a small way) for you to enjoy your outdoor experience? List three ways.
- At the time of your trip, did you feel that you were making a large impact on the forest around you? What do you think some of the long-term effects will be of numerous "consumers" in the area you visited?
- Projects such as the natural reserves in the Nilgiri Hills in South Asia include reforestation attempts; however, by establishing these reserves, they are attracting more nature-based tourists to the area. What are some efforts that should be made to lessen the impact of tourists in natural areas?
- Tourism also has many positive aspects, especially when local villagers earn much needed income and see the value of preservation. What can you do as a tourist to increase these positive effects?

IMPORTANT PLACES

The following places are featured in the chapter. Make sure you can locate all of them on a map. Blank outline maps can be found on the textbook's Web site: www.whfreeman.com/pulsipher4e. Also, to prepare for quizzes and exams, write a few important facts about each place in the space provided.

Physical Features

1. Arabian Sea
2. Bay of Bengal
3. Brahmaputra River
4. Deccan Plateau
5. Eastern Ghats
6. Ganga Plain
7. Ganga River
8. Himalayas
9. Hindu Kush
10. Indian Ocean
11. Indus River
12. Narmada River
13. Nilgiri Hills
14. Western Ghats

Regions/Countries/States/Provinces

15. Afghanistan
16. Ahraura
17. Bangladesh
18. Bhutan
19. Gujarat
20. India
21. Joypur

22. Kashmir

23. Kerala

24. Maldives

25. Nepal

26. Pakistan

27. Punjab

28. Sri Lanka

29. Uttar Pradesh

30. West Bengal

Cities/Urban Areas

31. Chennai (Madras)

32. Colombo

33. Delhi/New Delhi

34. Dhaka

35. Islamabad

36. Kabul

37. Kathmandu

38. Kolkata (Calcutta)

39. Mumbai (Bombay)

40. Thimphu

41. Varanasi (Benares)

MAPPING EXERCISES

The following are three mapping exercises to improve your knowledge of the location of places, underscore why they are important, and clarify how they relate to each other. Some questions will ask you to locate places, compare maps, or fill in data; others will test your understanding of *why* you were asked to map the features that you did. Use the blank outline maps at the end of the chapter to complete these exercises. Additional blank outline maps can be found on the textbook's Web site: www.whfreeman.com/pulsipher4e.

1. Relationship between landforms and population

South Asia has some of the most spectacular landforms and important rivers on earth. Some of these physical features make the land more habitable than others.

- From the map of the region (Figure 8.1), draw the outline, and label the Deccan Plateau.
- Draw (with ^^^) and label the location of these mountain ranges: Eastern Ghats, Western Ghats, Himalayas, and Hindu Kush (Figure 8.1).
- Draw a heavy blue line over these rivers: Brahmaputra, Ganga, Indus, and Narmada (Figure 8.1). Label each of these rivers.
- Shade the areas that have over 1300 persons per square mile (over 500 people per square kilometer) (Figure 8.13).

Questions

a. Based on the information you mapped, is population concentrated in specific locations (i.e., coastal or interior, along rivers or mountain ranges, etc.)? If so, where?
b. What is the general relationship between population and each of the different landforms: plateaus, mountains, and rivers? Explain the reasons for each.

2. How does climate affect people's lives and location?

This region has a great variety of regional climates, ranging from tropical monsoon to deserts. These physical conditions can have major impacts on living conditions.

- Shade in the general location of temperate climates (in green) and the arid and semiarid climates (in orange) (Figure 8.5).
- Use a hatch pattern (///) to show where population density is over 1300 persons per square mile (over 500 people per square kilometer) (Figure 8.13).
- Use the opposite hatch pattern (\\\) to show the areas with the heaviest monsoon rains (Figure 8.4).

Questions

a. Based on the information you mapped, what is the general relationship between population and each of the two climate types? Why is this the case for each?
b. Take a look at the areas where the hatch patterns cross (high population density *and* heavy monsoon rains). List at least three positive and three negative effects of monsoon rains on the people in these densely populated areas.

3. Female literacy and population growth rates

The overall status of women in South Asia is low; however, their relative well-being varies geographically. Freeing women from purdah encourages lower fertility rates and allows women to improve their own educational attainment and overall well-being, as well as that of their children.

- Using the table of human well-being rankings (Table 8.1), shade in each country's female literacy rate from light to dark using these categories: 0-30 percent; 31-50 percent; and 51-100 percent.
- Referring to Figure 8.15 (declines in total fertility rates), use a graduated symbol (from small to large circles, for example) to map the 2005 total fertility rate for the countries provided, using these categories: 0-2; 2.1-4; and 4.1-5.
- Finally, write in the number for the Gender Development Index (GDI) (Table 8.1).

Questions

a. Analyzing the information you mapped, describe what spatial pattern(s) you see.
b. Does population growth (total fertility rate) generally increase or decrease with education (female literacy rate)?
c. Discuss how GDI relates to education (literacy rate) and population growth (total fertility rate).
d. Do you see any anomalies (e.g., high literacy/high growth rate or low literacy/low growth rate)? Why might this be the case?

SAMPLE EXAM QUESTIONS

The following are sample questions to help you review for an exam. Answers are found in the back of this workbook.

1. What process created the Himalayan Mountains?
   A. Accumulation of magma surfacing along the boundaries of the Indian Plate
   B. The collision of the African Plate with the Indian Plate
   C. The collision of the Indian-Australian Plate with the Eurasian Plate
   D. Accumulation of magma surfacing through breaches in the Eurasian Plate

2. Among the following British actions in South Asia, which was beneficial to the region's economic development during and after the colonial period?
   A. Prohibiting tariffs on imported manufactures
   B. Constructing an expansive railroad network
   C. Preserving native languages to improve regional trade
   D. Importing British products to discourage industrial growth

3. Which of the following occurred as a result of the 1947 partition of British India?
   A. Muslims migrated from India to Pakistan; Hindus migrated from Pakistan to India.
   B. Muslims migrated from Pakistan to India; Hindus migrated from India to Pakistan.
   C. Muslims migrated from India to Sri Lanka; Hindus migrated from Sri Lanka to India.
   D. Muslims migrated from Sri Lanka to India; Hindus migrated from India to Sri Lanka.

4. Which of the following provides the least plausible explanation for the continued population boom in South Asia?
   A. Huge number of people in early reproductive years
   B. High literacy rates among women and men
   C. Relatively high infant mortality rates
   D. Preference for male children

5. Education affects population growth by encouraging which one of the following?
   A. A preference for male children
   B. The use of contraceptives
   C. Women to be viewed as an economic liability
   D. Men to desire larger families

6. Which of the following statements describes the practice known as *purdah*?
   A. Receiving payment from the family of a daughter's husband
   B. Concealing women from non-family members
   C. Killing of a wife who possessed an insufficient dowry
   D. Segregating people according to their occupation

7. Which of the following does NOT characterize any aspect of the green revolution in South Asia?
   A. Greatly boosted grain harvests
   B. Improved the mechanization of agriculture
   C. Benefits accrued to poorer farmers
   D. Introduced new seeds and fertilizers

8. Which of the following statements captures the idea of *microcredit* that has developed in South Asia?
   A. Accepts development loans only from lenders in the developing world
   B. Forgives all but a meager 5 percent of foreign debt owed by countries
   C. Makes small low-interest loans available to poor entrepreneurs
   D. Offers collateral-free loans to private, locally owned companies

9. The relations of what two South Asian countries are strained by regional conflicts in the states of Punjab and Kashmir?
   A. Sri Lanka and Bangladesh
   B. Bangladesh and India
   C. India and Pakistan
   D. Pakistan and Bangladesh

10. The "Chipko movement" in India is part of a broader reaction against what pattern found in South Asia?
    A. The increasing presence of high-tech firms from the West (United States) drive domestic high-tech firms out of business.
    B. Rural resources are used by urban industries without considering the needs of local (rural) people.
    C. The government refuses to adopt policies that would allow the mechanization of agriculture.
    D. City governments urge rural people to move to their cities but fail to provide them with sufficient housing.

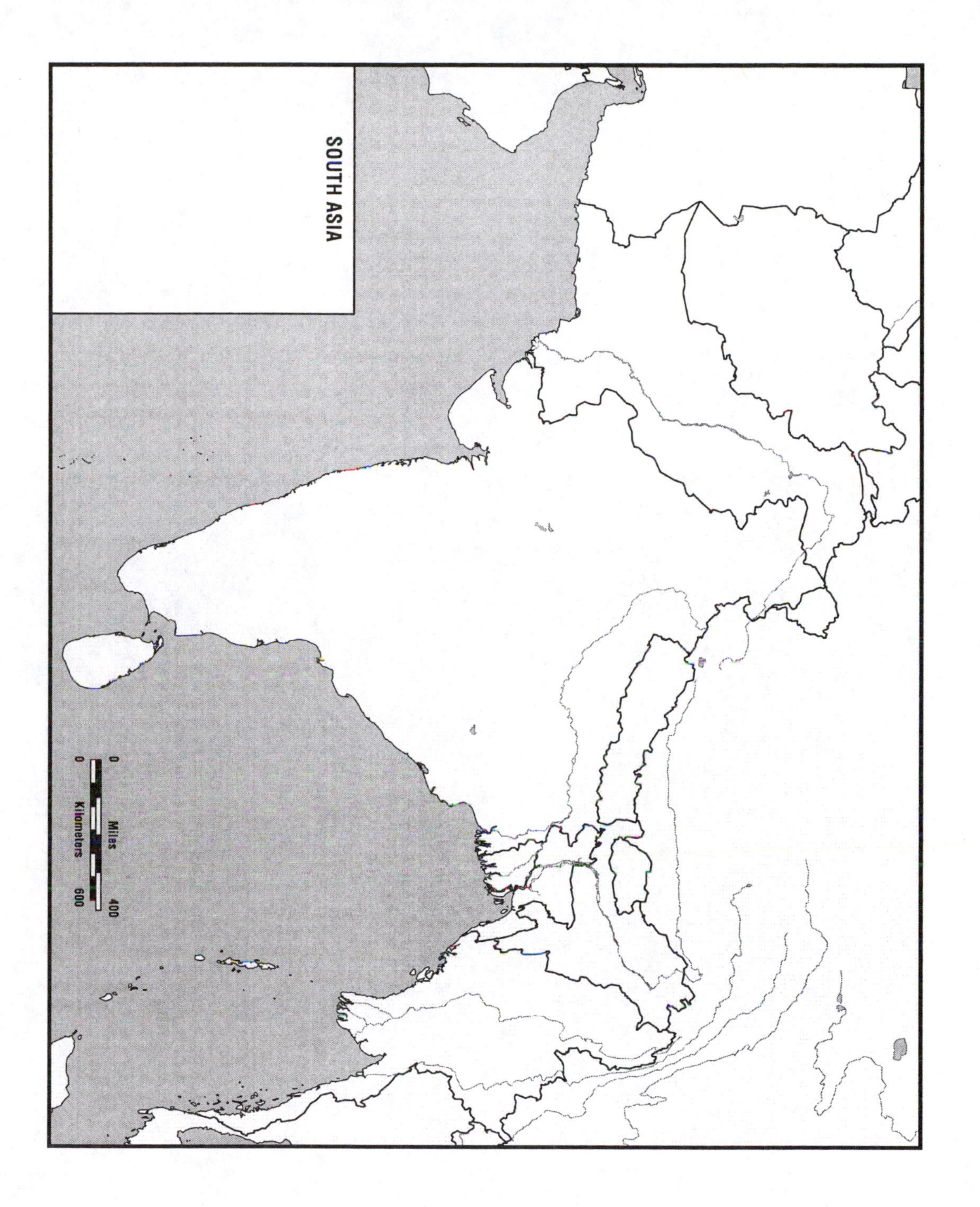
SOUTH ASIA
0 Miles 400
0 Kilometers 600

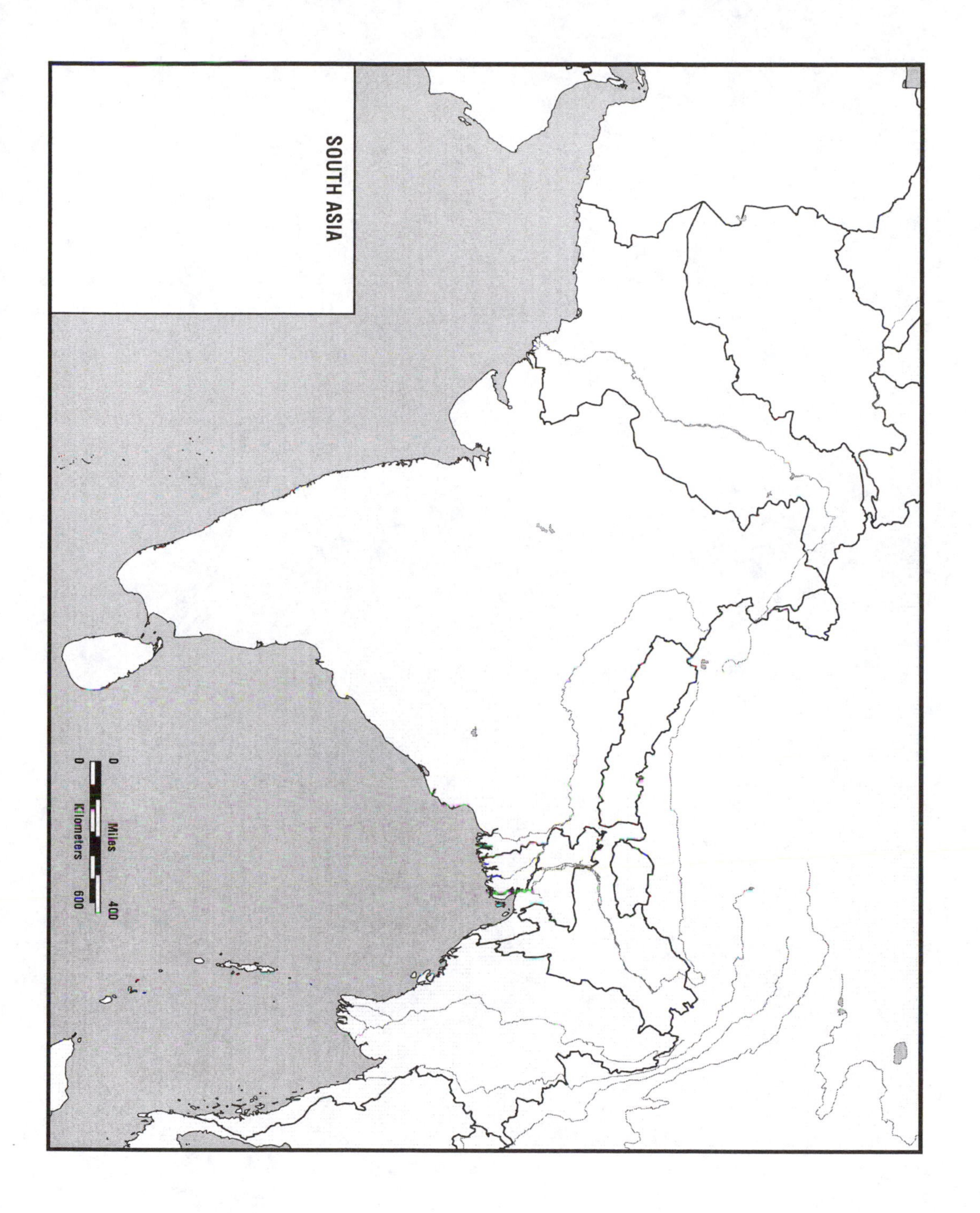
SOUTH ASIA
0 Miles 400
0 Kilometers 600

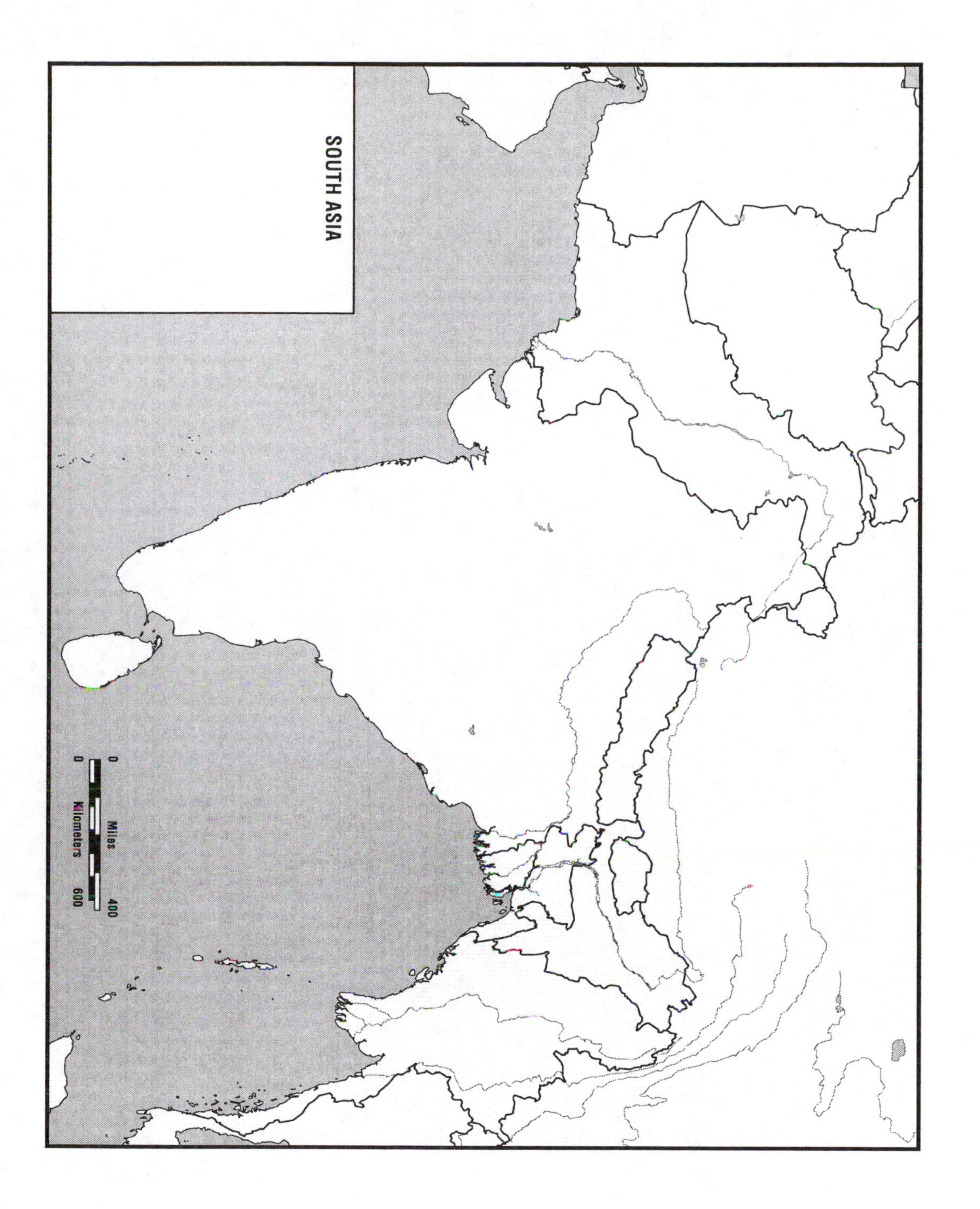
SOUTH ASIA
Miles
0
400
Kilometers
0
600

CHAPTER NINE
East Asia

LEARNING OBJECTIVES

After reading the chapter and working through this study guide, you should:

- Know how the various types of landforms were created in East Asia. Understand what drives the region's two main climate types. Know how landforms and climate affect agriculture as well as population distribution and density.
- Know how ancient Chinese forms of government and philosophy have influenced the development of the economy, culture, politics, and gender roles.
- Know why population is still growing in this region even though growth rates are declining. Understand how the growth of huge populations is putting strains on both human well-being and the environment. Know what is being done to attempt to control population growth.
- Know the differences between free-market and Communist economic systems. Know what market reforms have been enacted in this region. Know the positive and negative consequences of each type of reform.
- Know how gender attitudes are changing in the region. Understand family, work, and gender relationships in industrial areas of East Asia.
- Know how and why the Han Chinese majority dominates this region. Know the effects of discrimination on the numerous ethnic groups in this region.
- Understand the different environmental problems East Asia is facing. Know the causes and outcomes of these problems and how governments and citizens are trying to improve environmental health.

KEY TERMS

The following terms are in **bold** in the textbook. In the space next to the term (or on a separate sheet or flash cards), you can fill in the definitions for reference or quiz yourself for exam review. Definitions are found in the glossary of the textbook.

Ainu

alluvium

Asia-Pacific region

chaebol

Confucianism

Cultural Revolution

economic and technology development zones (ETDZs)

floating population

food security

food self-sufficiency

ger

Great Leap Forward

growth poles

hukou system

kaizen

loess

qanats

regional self-sufficiency

responsibility system

shogun

soft power

special economic zones (SEZs)

state-aided market economy

tsunami

typhoon

wet rice cultivation

yurt

REVIEW QUESTIONS

The following review questions are related directly to the textbook material. These questions can be used to help you prepare for an exam or you may want to read through the questions before you begin reading the textbook, quizzing yourself after you complete each section.

The Geographic Setting

1. There are few flat surfaces in East Asia. Those that are flat are often either too cold or dry for human use. What adaptations do East Asians make to be able to use this land?
2. Describe the characteristics of the four steps of East Asia's landforms. Discuss human habitation in each of the four steps; understand the reasons for very densely populated areas, as well as sparsely populated areas.
3. Compare the climate, vegetation, and population of the dry continental western zone and the monsoon east.
4. What are some positive and negative effects that Confucianism had on political, economic, and social life? Why was the industrial change slow during this time?
5. Know some of the differences between the Kuomintang (KMT) and the Chinese Communist party (CCP). Who backed them, who were their leaders, and what were their goals, philosophies, and impacts?
6. Why did people want a revolution in China? How did it affect agriculture, infrastructure, and gender roles?
7. What factors contributed to Japan's transformation into a democratic nation and a giant in the global economy?
8. What are the negative implications of China's large population? How might health issues like HIV-AIDS affect populations?

Current Geographic Issues

9. How did Japan overcome isolation and destruction from World War II to become one of the world's wealthiest and most influential nations?
10. What are some of the differences between the economies of capitalist countries and Communist countries? What are some of the positive and negative aspects of each?
11. Did communal land reform work in China? Why or why not? What were the benefits and drawbacks of this system?
12. Why was China's policy of regional self-sufficiency encouraged? Discuss its successes. Have market reforms and responsibility systems been more successful? Why or why not?
13. What are some important economic reforms in this region? Who are they affecting? How are they affecting the environment? How are they affecting the distribution of wealth and standards of living?

14. What effects will a growing population have on the economy, environment, and human well-being in China? What are some of the positive and negative effects China's one-child-per-family policy is having? Why does China's population continue to grow, even though couples are limited to one child?
15. What long-term effects will population control have on women in China?
16. What are the roles of most urban women? Why don't women *want* to work in jobs with full responsibilities? What group of people is dominant in challenging the roles of urban family structure? Why?
17. How are minority groups treated in East Asia? Why are Han settlers being sent to regions that have large numbers of minorities? What are the effects of this resettlement?
18. What are some of the problems with air quality in East Asia? How are these related to urbanization and population density?
19. What are the problems with water in East Asia, and how are humans making things worse? What are the effects of too much water? Too little water? How are people reacting to and dealing with these problems?
20. Although there have been many improvements in quality of life, including health care, water quality, and literacy, what has happened to cause quality of life to decline recently?

CRITICAL THINKING EXERCISES

The following are "what if"-type questions that illustrate concepts you are learning from the textbook. These questions ask you to apply the ideas and principles you learned from the textbook to new situations.

1. Confucian ideology and society

The Confucian ideology penetrated all aspects of Chinese society and deeply affected its social, economic, and political geography.

- A woman was confined to the domestic spaces of home and almost always placed under the authority of others: her parents, her husband, or her son. What might life have been like for women? What might life have been like for the men in the seat of authority?
- What might life have been like for the masses at the base of the hierarchical pyramid?
- Confucian values were used to organize the state as well as society. Do you think people were better off with this ruling tradition? What would have been some other alternatives?

2. The communal way of living

When the Communist party first came to power in China, it undertook land reform, which resulted in banding small landholders into cooperatives, then eventually into full-scale communes in an attempt to improve agricultural production.

- What were the benefits and drawbacks of the commune system? Do you think it did more good or more harm overall?
- Communes took over all aspects of life, including political organization, industry, health care, and education. What would be some of the benefits and drawbacks in *your life* if you were forced to or chose to live communally?
- What would some obstacles be to instituting this type of system in the United States?

3. Would you invest in China?

China has begun to pursue a more efficient and market-oriented economy. It has become a participant in the global economy as a significant producer of manufactured goods, with a market of more than 1 billion customers.

- What factors would discourage you from investing or locating a business in China?
- What factors would encourage you to invest or locate a business in China?
- Based on the factors you've listed, would you invest or locate a business in China? What factor(s) really made the decision for you?
- Identify five changes that would have to take place in this region to make investing in China more attractive. Be sure to consider social, economic, political, and environmental factors.

4. Effects of population control on the extended family

In an attempt to reduce population growth in a country with already over one-fifth of the world's population, most Chinese couples are limited by law to one child. Within two generations, the kinship categories of sibling, cousin, aunt and uncle, and sister- and brother-in-law disappeared from families that complied with the policy.

- Because of the great value people put on an extended family in China, how do you think this will affect family life? How will it affect child and elder care?
- How would this policy have affected *you* and *your family* if it were in place when your parents or grandparents were having children? What would it be like to have no siblings, cousins, or aunts and uncles?
- If you have siblings, consider who was first born in *your* set of siblings. Under a one-child policy, would you have been born? What might have happened in your family if the first born was a female? What if the first born was a male?
- How about when you have/had children? How would you feel about being legally limited to one child?

5. Tolerance of ethnic diversity in the Xinjiang Uygur Autonomous Region

Beijing's central authority has been under increasing challenge from Muslim separatists in the Xinjiang Uygur Autonomous Region of western China. The conflict has resulted in a significant loss of life and human rights violations against the Uygurs and violent incidents by separatists.

- Explore the following Web sites and identify examples of human rights violations in Xinjiang. Also, make a list of the four most commonly mentioned abuses.
  - www.amnestyusa.org – Amnesty International (search Uygur, Uyghur, or Uighur)

- www.uyghuramerican.org - The Uyghur American Association
- www.uyghurcongress.org - World Uyghur Congress
- www.uygur.org - East Turkistan Information Center

- There have also been a number of violent incidents that have been attributed to independence activists. Using these same Web sites, identify examples of violent incidents attributed to Muslim separatists.
- Thus far, the independence movement in Xinjiang has failed to generate widespread support and remains too fractured to present a meaningful threat to Beijing's rule. Why do you think this is the case?
- Do you think that the increasingly savage suppression of Muslim protests will generate unity and coordination within the various separatist groups in Xinjiang?
- What are the stakes in this conflict for both sides? Why are the Uyghurs and the Beijing government in such conflict?

IMPORTANT PLACES

The following places are featured in the chapter. Make sure you can locate all of them on a map. Blank outline maps can be found on the textbook's Web site: www.whfreeman.com/pulsipher4e. Also, to prepare for quizzes and exams, write a few important facts about each place in the space provided.

Physical Features

1. Chang Jiang (Yangtze)

2. East China Sea

3. Gobi Desert

4. Himalayas

5. Huang He (Yellow River)

6. Junggar Basin

7. Mekong River

8. Mongolian Plateau

9. Mount Fuji

10. North China Plain

11. Nu (Salween River)

12. Ordos Desert

13. Pacific Ocean

14. Pacific Ring of Fire

15. Plateau of Tibet (Xizang-Qinghai Plateau or Northern Plateau)

16. Qaidam Basin

17. Sea of Japan

18. Sichuan Basin

19. South China Sea

20. Taklimakan Desert

21. Tarim Basin

22. Zhu Jiang (Pearl River)

23. Yellow Sea

24. Yunnan Guizhou Plateau

Regions/Countries/States/Provinces

25. China

26. China's Far Northeast

27. Guangdong

28. Hong Kong

29. Inner Mongolia (Nei Monggol)

30. Japan

31. Macau

32. Mongolia

33. North Korea

34. South Korea

35. Taiwan

36. Tibet (Xizang)

37. Xinjiang Uygur

Cities/Urban Areas
38. Beijing

39. Guangzhou

40. Kyoto

41. Osaka

42. Pyongyang

43. Seoul

44. Shanghai

45. Taipei

46. Tokyo

47. Ulan Bator

MAPPING EXERCISES

The following are three mapping exercises to improve your knowledge of the location of places, underscore why they are important, and clarify how they relate to each other. Some questions will ask you to locate places, compare maps, or fill in data; others will test your understanding of *why* you were asked to map the features that you did. Use the blank outline maps at the end of the chapter to complete these exercises. Additional blank outline maps can be found on the textbook's Web site: www.whfreeman.com/pulsipher4e.

1. Population and landforms

Of the few flat surfaces in the rugged landscapes of East Asia, many are too cold or dry to be useful to humans.

- On a blank map of East Asia, shade the areas (in red) that have over 1300 persons per square mile (over 500 people per square kilometer) (Figure 9.14).
- From Figure 9.1 (regional map), shade (in tan) and label the location of the Gobi, Ordos, and Taklimakan deserts.
- Trace with a thick blue line and label the following rivers: Chang Jiang (Yangtze), Huang He (Yellow River), Mekong River, Nu (Salween), and Zhu Jiang (Pearl River).
- Draw (with ^^^) and label the location of the Himalayas.
- Draw a hatch pattern over and label the Plateau of Tibet and Yunnan Guizhou Plateau.

<u>Questions</u>

a. Is population concentrated near any particular landform(s)?
b. Which landforms are associated with low/no population concentrations?
c. In cases where population is not located near a water source, how might people adapt to make agriculture productive?
d. In cases where population is located near rugged terrain, what adaptations might people make to create space for agriculture?

2. Economic growth, SEZs, and ETDZs

Special economic zones (SEZs) and economic and technology development zones (ETDZs) are central to China's new market reforms and have rapidly opened the economy to international trade.

- Using Figure 9.22 (map of foreign investment), on the blank map of China and its provinces, draw a blue square for each SEZ and a red dot for each ETDZ.
- Using Figure 9.21 (map of rural-urban GDP disparities), shade the provinces in green that have over 9000 yuan GDP per capita (use yellow for the "upper middle" category and orange for the "high" category).

<u>Questions</u>

a. What type of general relationship did you expect to find between the SEZs/ETDZs and GDP per capita? Explain why SEZs/ETDZs might affect GDP per capita.
b. Are there provinces with SEZs/ETDZs that don't have an upper-middle or high GDP? Provide at least two reasons why do you think this is the case.
c. Assume that SEZs/ETDZs are major growth poles (drawing more investment and migration) and examine the map of population density (Figure 9.14) and the map of rural-to-urban migration (Figure 9.3).
    - What will happen to those provinces with SEZs/ETDZs that already have high population densities (consider the environment, infrastructure, and resulting living conditions)?
    - Why do you think the government has chosen to put SEZs/ETDZs in areas with population densities? What effects might SEZs/ETDZs have on these areas?

3. The Three Gorges Dam

Although the Three Gorges Dam is expected to save millions of lives and much property, the 370-mile long reservoir it creates will drown 62,000 acres of farmland, 13 major cities, 140 large towns, hundreds of small villages, and 1600 factories, displacing over 1.9 million people.

- On the blank map of the Central China subregion, draw the Chang Jiang with a thick blue line.
- Using the map of population density (Figure 9.14), shade in areas that have over 1300 people per square mile (over 500 people per square kilometer).
- Using the map of foreign investment (Figure 9.22), draw a red circle for each ETDZ and a blue square for each SEZ.
- Using the maps on the University of Hong Kong's Civil Engineering Computer Aided Learning (CIVCAL) Web site (civcal.media.hku.hk/threegorges/Default.htm), draw in the location of the dam.
- Also, using the maps on the CIVCAL Web site, draw the outline, showing the extent of the reservoir that will be created by the Three Gorges Dam.

Questions

a. Based on the outline of the reservoir, what cities or urban areas might be affected and how?

b. Will this dam negatively affect any of the ETDZs or SEZs by flooding them or reducing the amount of water flowing in the area? Which ones?

c. Will the dam/reservoir bring any ETDZs or SEZs *more* business? Which ones and why?

d. Using the University of Hong Kong's Web site, identify two positive effects the dam/reservoir might have on each of the following: society, politics, and the environment.

e. Evaluate the University of Hong Kong's Web site: What type of information is missing from this site? Find at least two other critical Web sites and identify at least five negative effects the dam/reservoir might have on: society, politics, and the environment.

SAMPLE EXAM QUESTIONS

The following are sample questions to help you review for an exam. Answers are found in the back of this workbook.

1. Which of the following occurs during East Asia's winter monsoon?
   A. Warm moist air sweeps east and south through East Asia.
   B. Cold moist air blows west and north through East Asia.
   C. Cold dry air blows east and south through East Asia.
   D. Warm moist air sweeps west and north through East Asia.

2. Which of the following accords with Confucian ideology?
   A. A male should hold the position of ultimate authority.
   B. The matriarchal family is the best organizational model for the state.
   C. The military general is the source of all order and civilization.
   D. Imperial officials are accountable to common people.

3. Which of the following was a main factor causing Japan's recent economic problems?
   A. Sudden deregulation and removal of tariffs
   B. Sudden decline in wages and lack of health benefits
   C. Increased foreign investment from Southeast Asia pulling money away from Japan's economy
   D. Corrupt relationships grew between officials in government and industry.

4. Which of the following characterizes the land reform strategy enacted by the Chinese Communist party when it first came to power in 1949?
   A. Formed private agricultural corporations out of large tracts of land unused by the aristocracy
   B. Turned formerly agricultural lands over to herders
   C. Divided large estates of landlords among landless farmers
   D. Sold state land holdings to retired soldiers and unemployed merchants

5. Which of the following characterizes China's regional development policy during the first two-and-a-half decades of Communist leadership?
   A. Each region was to identify its comparative advantage and target its resources to develop that sector.
   B. Coastal regions would emphasize foreign-funded, export-oriented industrial development while interior regions would emphasize domestically funded industry.
   C. Each region was encouraged to develop agricultural and industrial sectors independent of other regions.
   D. Regions were paired up and encouraged to coordinate the development of a symbiotic relationship between their economies.

6. Which of the following was among China's internal reform strategies pursued in the late 1970s?
   A. Regional specialization was encouraged.
   B. Economic decision making was centralized.
   C. Competition was discouraged.
   D. Tariffs were introduced.

7. The term *floating population* represents which group of people in China?
   A. Skilled workers who, during periods of rapid job creation, hop from one job to another in search of the best opportunity
   B. Jobless or underemployed rural people who have migrated to the cities without official permission
   C. Young women who have left the cities without permission to find higher status work in the labor deficient countryside
   D. Foreign companies with special permission from the government to move their operations, including workers, in pursuit of favorable conditions

8. According to the textbook, what is the main reason the Chinese government has moved hundreds of Han settlers into the far northwest zones of China?
   A. To provide military protection against Central Asian terrorists
   B. To encourage the Han to adopt the Islamic culture
   C. To prevent the emergence of a cultural bridge between minorities and mainstream Chinese
   D. To weaken the power of the indigenous minority groups

9. Which of the following does NOT correlate with the geographic pattern of air pollution in China?
   A. Industrial development
   B. Urban areas
   C. Use of hydroelectric energy
   D. Population density

10. Under Communist rule, which of the following has NOT been improved in China?
   A. Literacy training
   B. Health care delivery
   C. Air quality
   D. Poverty rate

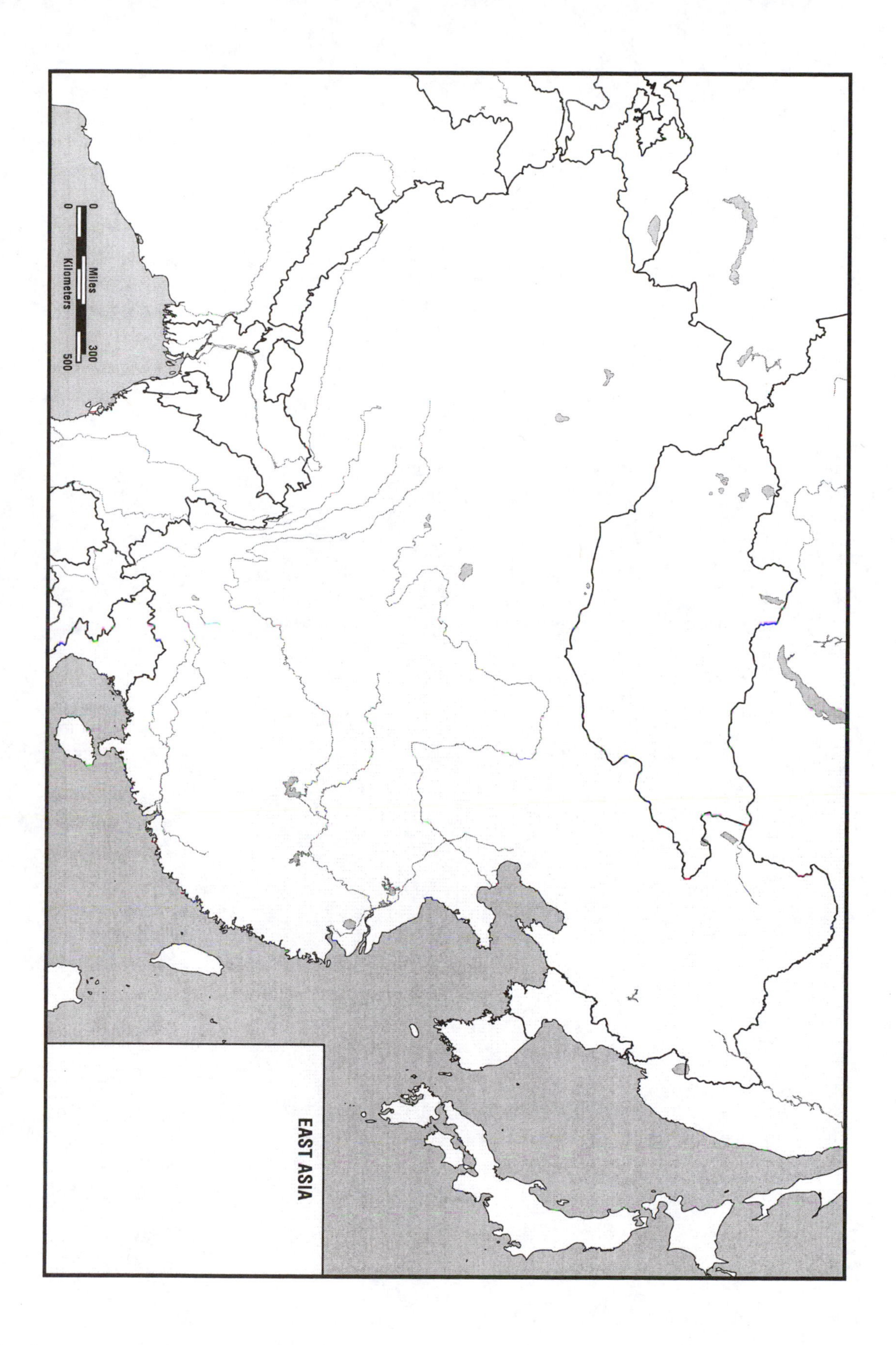
EAST ASIA
Miles
0
300
Kilometers
0
500

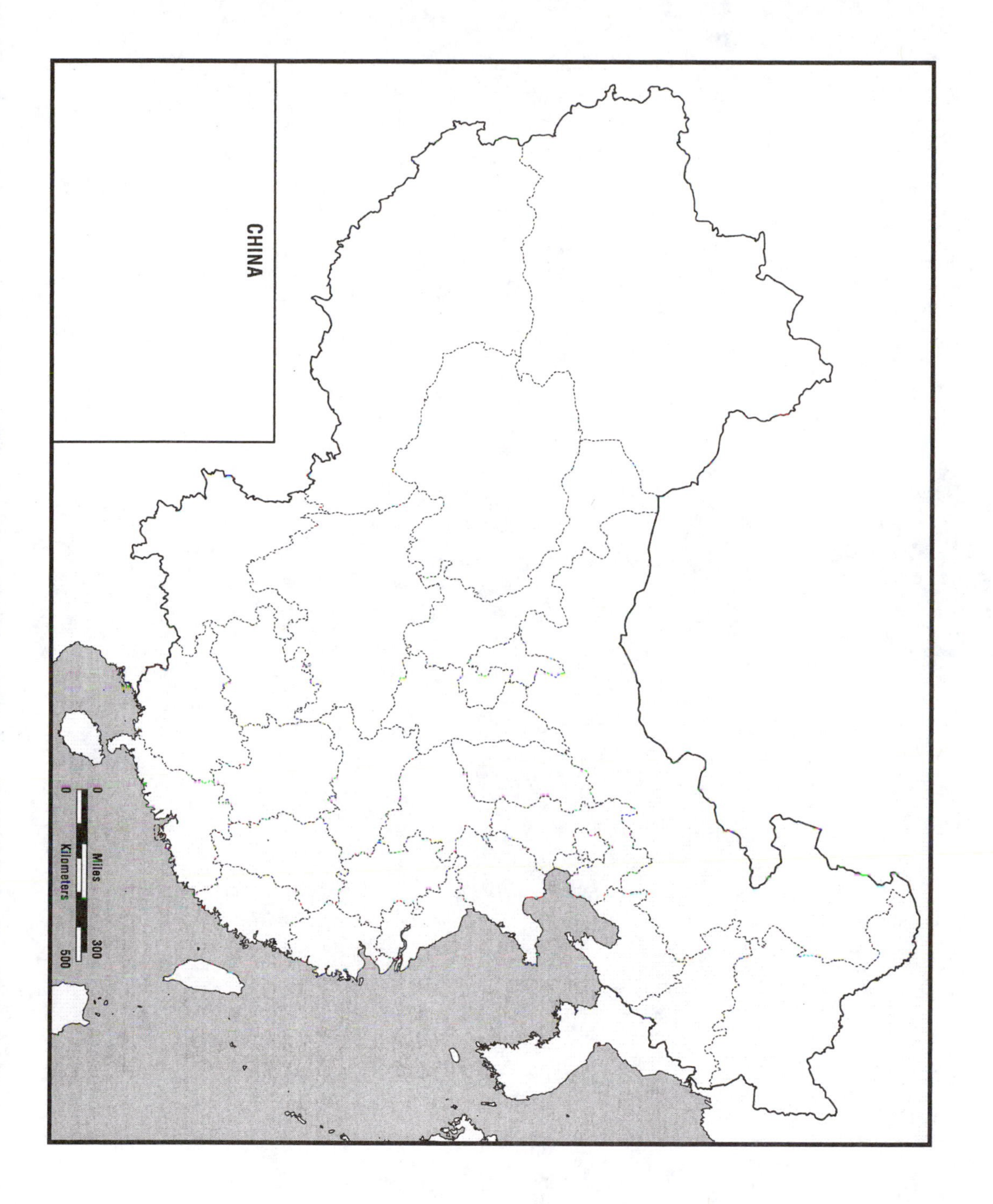
CHINA
0 300
Miles
0 500
Kilometers

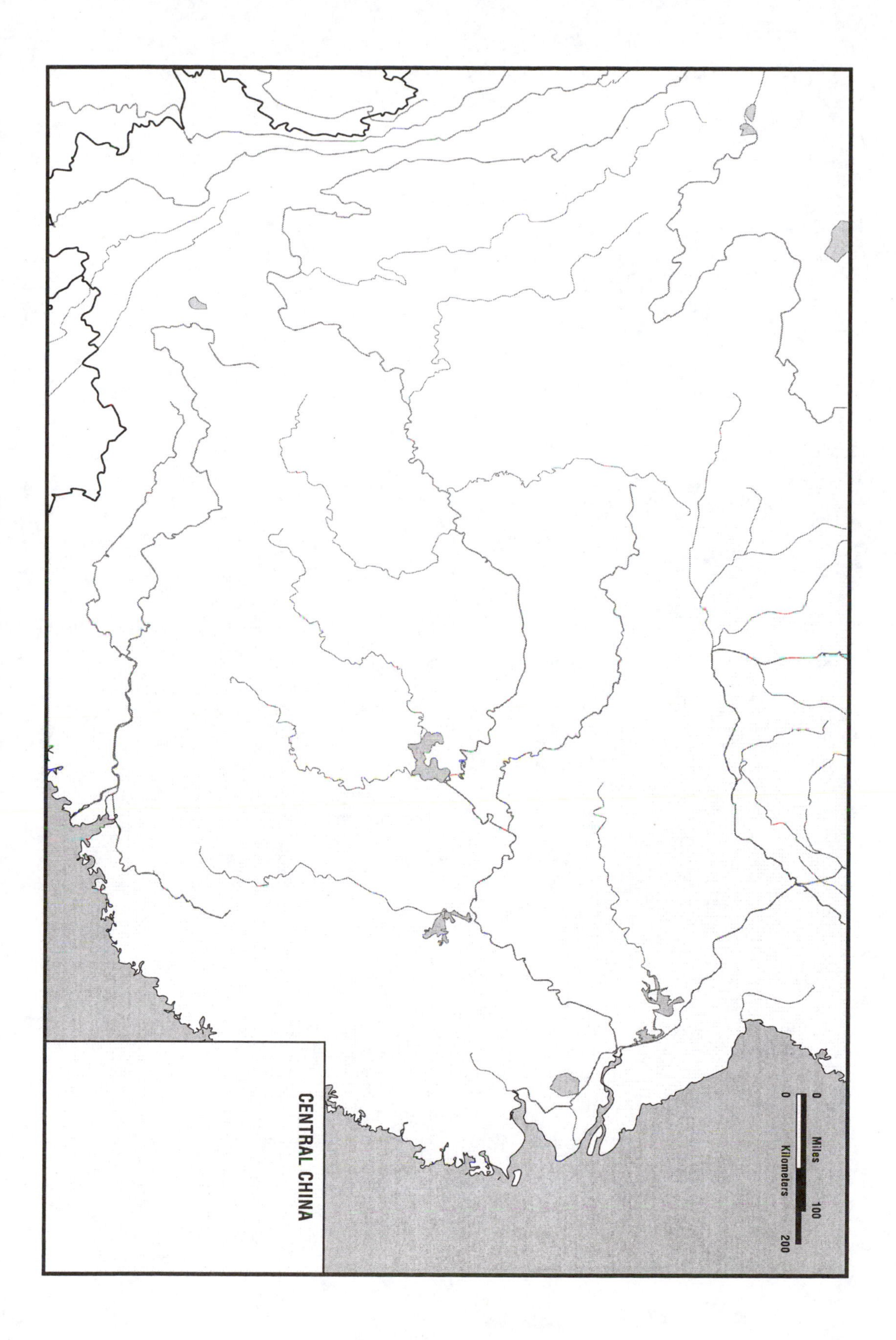
CENTRAL CHINA
0
0
Miles
Kilometers
100
200

CHAPTER TEN
Southeast Asia

LEARNING OBJECTIVES

After reading the chapter and working through this study guide, you should:

- Know how this region of islands and peninsulas has managed to use its physical location to create strong linkages with the global economy.
- Know how the legacy of European colonization and, more recently, global assembly plants, left an impact on the cultural landscape, settlement patterns, economy, environment, and religion.
- Understand how the land is able to support twice the population of the United States, but with negative consequences on the environment.
- Understand why some countries in the region are mostly agricultural and just beginning to modernize, while others are industrialized and have large modern cities. Understand the relationship between different livelihoods and human well-being in both rural and urban contexts.
- Identify those economic activities that have contributed to the region's rapid growth rate and those that are exploitative of people and the environment.
- Explain how so many distinct cultural and religious groups have lived side-by-side for long periods in Southeast Asia.
- Know why the status of women in Southeast Asian society is generally better than the patriarchal structure might suggest. Discuss reasons for the relatively high status of women in this region.

KEY TERMS

The following terms are in **bold** in the textbook. In the space next to the term (or on a separate sheet or flash cards), you can fill in the definitions for reference or quiz yourself for exam review. Definitions are found in the glossary of the textbook.

archipelago

ASEAN Free Trade Association (AFTA)

Association of Southeast Asian Nations (ASEAN)

Australo-Melanesians

Austronesians

crony capitalism

cultural complexity

cultural pluralism

detritus

doi moi

Export Processing Zones (EPZs)

extraregional migration

feminization of labor

foreign exchange

growth triangles

old-growth forest

Pancasila

pull factors

push factors

resettlement schemes

sex tourism

wet (paddy) rice cultivation

REVIEW QUESTIONS

The following review questions are related directly to the textbook material. These questions can be used to help you prepare for an exam or you may want to read through the questions before you begin reading the textbook, quizzing yourself after you complete each section.

The Current Geographic Setting

1. Explain how plate tectonics play a role in creating the conditions for active volcanism in the Philippines and Indonesia.
2. Explain how the monsoons and ITCZ bring rain to Southeast Asia most of the year.
3. The human settlement of Southeast Asia is ancient. Describe the two prehistoric waves of people into the region and the landforms that facilitated this settlement. Where have settlers come from in the last 2000 years? What has prompted these newer waves of migration?
4. Describe the conditions under which Europe, the United States, and Japan colonized various parts of Southeast Asia at different times. When was independence achieved in the various colonies and under what conditions? Why do some observers say that Southeast Asia is still a dependent region of the world despite the demise of colonization?
5. Where has population grown most rapidly, and where are densities high? Where is population relatively sparse? How is population growth correlated with wealth and poverty in the region?
6. How do religious and cultural traditions promote the transmission of HIV-AIDS?

Current Geographic Issues

7. List and describe three different agricultural systems that are found in the region. Identify those that are dying out and those that are gaining in importance. Explain the reasons for these changes.
8. How did national governments use import substitution industries as a successful investment method?
9. How are export processing zones (EPZs) similar to or different from the North American Free Trade Agreement (NAFTA) or the European Union (EU)?
10. What is the role of sex tourism in development? Why is this detrimental to overall development?
11. What factors led to the economic crises of the late 1990s? What procedures or practices helped the countries pull out of the crises?
12. What countries have benefited the least from industrialization? Which have suffered the most when economic downturns occur? Why is this the case?
13. Why are the Overseas (ethnic) Chinese criticized in the region?
14. What major religions are important in the region? To some extent, different religions are associated with different ethnic groups and parts of the region. Describe the alignments for Animism (traditional), Buddhism, Islam, Hinduism, and Christianity.
15. Southeast Asia is a region of significant religious pluralism. As a consequence of exposure to each other, some religious practices have experienced change. Give two case studies and explain the rationale for religions to adopt new practices.
16. What has been the impact of modernization on families? What are the benefits of these changes? What are the challenges?

17. Women play important and powerful roles in Southeast Asia. Discuss at least three aspects of society and family life that illustrate this fact.
18. How does rural-to-urban migration affect agricultural production? How can countries solve these problems? List three push factors and three pull factors of migration. How do resettlement schemes solve some problems while exacerbating others?
19. What practice accounts for the greatest depletion of the rainforest? Does this surprise you? Why or why not? Deforestation leads to what other general environmental problems?
20. How do the socialist countries of Vietnam, Cambodia, and Laos compare in human well-being achievements with other countries in the region? Why is Vietnam doing better on the HDI scale than Laos and Cambodia?

CRITICAL THINKING EXERCISES

The following are "what if"-type questions that illustrate concepts you are learning from the textbook. These questions ask you to apply the ideas and principles you learned from the textbook to new situations.

1. Following in Singapore's footsteps!

Hong Kong, Singapore, Taiwan, and South Korea are successful newly industrialized countries (NICs) in Asia. As we observe other countries in Southeast Asia, we might ask, "Which ones will be the next to successfully industrialize?"

- Which of the following countries - Thailand, Malaysia, Indonesia, the Philippines, Laos, Cambodia, and Vietnam - are likely to be added to the list of countries that have made the transition from low-wage manufacturing to sophisticated, mature, and diversified economic activity?
- Justify your choice(s) and explain why you believe your choice(s) will follow in Singapore's footsteps.

2. Burma and Thailand: How can two neighbors be so different?

Burma and Thailand share a common border and part of the Malay Peninsula, yet their historical paths and current status are vastly different.

- Compare and contrast the historical and current political, social, and economic characteristics of these two neighbors.
- What expectations do you have for each country's development during the twenty-first century?
- Will these two neighbors ever share common development goals? Defend your response.

3. Vietnam: Two generations' perspectives

Many people born during the last 25 years of the twentieth century have little or no memory of the Vietnam conflict.

- Develop three questions that are of interest to you and interview someone you know (perhaps your father or uncle, maybe even your aunt or mother) who served in Vietnam during the Vietnam conflict.
- Ideally, see if you can find Vietnamese people who have now immigrated to the United States and remember the Vietnam conflict. Ask them the same questions.
- Now interview two of your fellow students and ask them the same questions. If possible, find some Vietnamese or Vietnamese-American students on campus and interview them.
- Draw some conclusions about different generations' and different nationalities' attitudes about Vietnam and about the conflict.

4. ASEAN's growing membership

ASEAN seeks to form a regional trade membership by using the North American Free Trade Agreement (NAFTA) and the European Union (EU) as models.

- What are positive and negative aspects of membership in ASEAN for the current members?
- Why does ASEAN want to bring in the poorer countries of Southeast Asia? How can current ASEAN members benefit by bringing in these poorer members? How can the poorer countries benefit?

5. Coping with fragmentation

Several Southeast Asian countries are fragmented spatially. For instance, Indonesia is spread across an archipelago that includes 17,000 islands.

- What challenges do the governments of these fragmented countries face in trying to establish a strong centralized power?
- What economic, political, and social challenges arise from this fragmentation?
- How do people whose geographical separation is often associated with weak nationalistic ties develop a sense of commitment to one national government?

IMPORTANT PLACES

The following places are featured in the chapter. Make sure you can locate all of them on a map. Blank outline maps can be found on the textbook's Web site: www.whfreeman.com/pulsipher4e. Also, to prepare for quizzes and exams, write a few important facts about each place in the space provided.

Physical Features

1. Bali

2. Black River

3. Borneo

4. Chao Phraya River

5. Gulf of Thailand

6. Irrawaddy River

7. Java

8. Lesser Sunda Islands

9. Luzon

10. Madura

11. Malay Peninsula

12. Mekong River

13. Mindanao

14. Molucca Islands

15. Mount Pinatubo

16. New Guinea

17. Red River

18. Salween River

19. South China Sea

20. Strait of Malacca

21. Sulawesi (Celebes Islands)

22. Sumatra

23. Timor Island

Regions/Countries/States/Provinces

24. Brunei

25. Burma (Myanmar)

26. Cambodia

27. East Timor (Timor-Leste)

28. Indochina

29. Indonesia

30. Kalimantan

31. Laos

32. Malaysia

33. Philippines

34. Sabah

35. Sarawak

36. Singapore

37. Thailand

38. Vietnam

39. West Papua

Cities/Urban Areas

40. Bandar Seri Begawan

41. Bandung

42. Bangkok

43. Chiang Mai

44. Hanoi

45. Ho Chi Minh City (Saigon)

46. Jakarta

47. Kuala Lumpur

48. Manila

49. Phnom Penh

50. Rangoon

51. Vientiane

MAPPING EXERCISES

The following are three mapping exercises to improve your knowledge of the location of places, underscore why they are important, and clarify how they relate to each other. Some questions will ask you to locate places, compare maps, or fill in data; others will test your understanding of *why* you were asked to map the features that you did. Use the blank outline maps at the end of the chapter to complete these exercises. Additional blank outline maps can be found on the textbook's Web site: www.whfreeman.com/pulsipher4e.

1. Tourism and indigenous peoples

Tourism is suggested as a good alternative for regional and local economic development in Southeast Asia. However, indigenous peoples may be adversely affected while they, at the same time, reap economic benefits.

- Using Figure 10.3 (map of indigenous groups), and especially the inset map, shade the indigenous groups' locations on a blank map of Southeast Asia.
- Draw in the approximate location of the Asian highway system, using Figure 10.20 (map of transportation infrastructure).
- Also using Figure 10.20, mark the location of world heritage sites with a black dot.
- Finally, use Figure 10.10 (population density map) to cross hatch (///) areas with population densities of 651 or more people per square mile (251 or more people per square kilometer).

Questions

a. Based on your map, as well as the reading of the first few pages of the chapter, the section entitled "Tourism Development," and Box 10.1 (Tourism Development in the Greater Mekong Basin), identify those areas where indigenous peoples will be highly affected by a developing tourist industry.

b. Suggest benefits to the indigenous peoples in the places you have identified. Make a judgment as to the sustainability of those benefits.
c. Suggest special challenges these indigenous people may face and discuss how their culture can be protected.

2. Agriculture and population density

Population density can often be related to economic lifestyles; even varying types of agricultural systems are often associated with different population densities.

- On a blank map of Southeast Asia, label all of the countries in the region.
- Using Figure 10.16 (map of agricultural patterns), shade intensive cropland.
- Using Figure 10.10 (map of population density), cross hatch (///) the areas with 651 or more people per square mile (251 or more people per square kilometer).

Questions

a. What is the relationship between intensity of agricultural pursuits and population density?
b. Are there any anomalies? If so, where are they? Explain them.

3. The diaspora of Southeast Asian women

Over 50 percent of the people who engage in extra regional migration are women. Many of these women move within the region as well as to other world regions to find work as maids.

- Using Table 10.4 (human well-being rankings), draw a graduated circle in each country to represent its GDP. Remember to make the smallest circle correspond to the lowest range of GDP.
- Again, using Table 10.4, select gray shades to shade female literacy for all countries. Remember to use the lightest gray for the lowest levels of literacy and the darkest shade for the highest level of literacy.

Questions

a. Refer to the "maid trade" map (Figure 10.30) and identify three countries in this region that appear to be major contributors to the "maid trade." Explain any relations you see between GDP and female literacy and countries' contribution to the "maid trade."
b. If no relationship is suggested by the map, then using your understanding of this region, suggest other reasons for a country's high "maid trade" numbers.

SAMPLE EXAM QUESTIONS

The following are sample questions to help you review for an exam. Answers are found in the back of this workbook.

1. According to the explanation in the textbook, how do periodic El Niño events help to create drought in Southeast Asia?
   A. Cooling ocean temperatures
   B. Warming ocean temperatures
   C. Lowering air pressure over land
   D. Lowering air pressure over water

2. According to the textbook, how did Thailand manage to maintain its independence during the European colonial era?
   A. Used diplomatic acumen to acquire protective services of the Soviet Union
   B. Poured resources into its military technologies that rivaled those of Europe
   C. Undertook a large-scale push to modernize itself in the European mold
   D. Assertively established its own colonies to serve as buffers around its territory

3. The collision of the Indian-Australian Plate and the Philippine Plate with the Eurasian Plate has contributed what sort of feature to the physical geographies of Indonesia and the Philippines?
   A. Rift valleys
   B. Trenches
   C. Volcanoes
   D. Coastal escarpments

4. Which of the following reflects the most common migration pattern within Southeast Asia?
   A. Moving from a large city to the countryside
   B. Moving from the island countries to those on the mainland
   C. Moving from a rural village to the country's largest city
   D. Moving from the mainland countries to those on the islands

5. According to the textbook, all EXCEPT which of the following are major factors in the spread of HIV-AIDS in Southeast Asia?
   A. Religious traditions restrict public sex education.
   B. Popular customs often permit a broad array of sexual practices.
   C. Fertility rates are decreasing rapidly.
   D. Many, not knowing they are infected, don't change their sexual practices.

6. Which of the following statements most accurately describes the cultural character of Southeast Asia?
   A. Many different groups have lived together for such a long time that differences between them are imperceptible.

B. Many different groups have lived together for a long time, yet have remained distinct.
C. Similar groups have lived together for a short time but have become more distinct since European colonialism.
D. Similar groups have lived together for a long time and are now essentially one large cultural group.

7. Which of the following descriptions most aptly describes the so-called Overseas Chinese in Southeast Asia?
   A. Chinese soldiers stationed at numerous naval and army bases along the Southeast Asian coastline
   B. Ethnic Chinese immigrants who have achieved economic success while living in Southeast Asia.
   C. Asian immigrants to Europe during colonialism who were assumed to be of the same ethnic group
   D. Chinese immigrants who form the Southeast Asian underclass working as domestic servants and street cleaners

8. In the textbook, which of the following points are made about families in Southeast Asia?
   A. Families are nearly always headed by the oldest surviving male.
   B. Men tend to serve as the family money managers.
   C. Newly married couples often live with the husband's parents.
   D. Wealth and power is passed down from father to oldest son.

9. Among the following forms of agriculture, which tends to be the LEAST sustainable over long periods in the physical environment (such as soil type) of Southeast Asia?
   A. Commercial farms
   B. Slash-and-burn farming
   C. Shifting cultivation
   D. Paddy cultivation

10. Which of the following is the most appropriate description of crony capitalism in Southeast Asia?
    A. Personal and familial connections among politicians, bankers, and entrepreneurs are used to create economic opportunities.
    B. Close ties formed among small groups of states allow them to jointly coordinate development projects.
    C. Private interests make investment decisions while state planners have responsibility for daily business operations.
    D. Companies invest outside of their areas of specialization in order to control all stages of the production process.

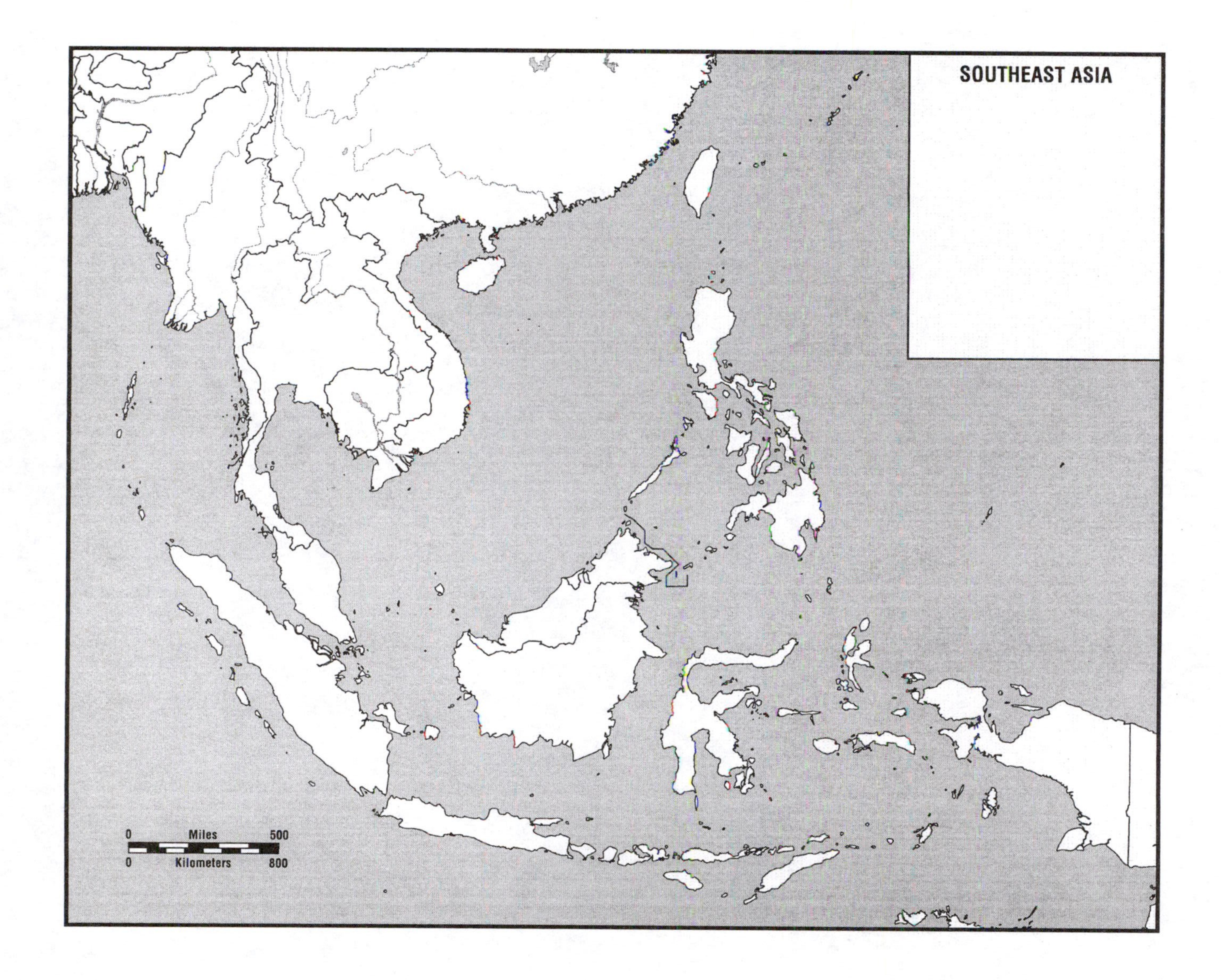
SOUTHEAST ASIA
0 Miles 500
0 Kilometers 800

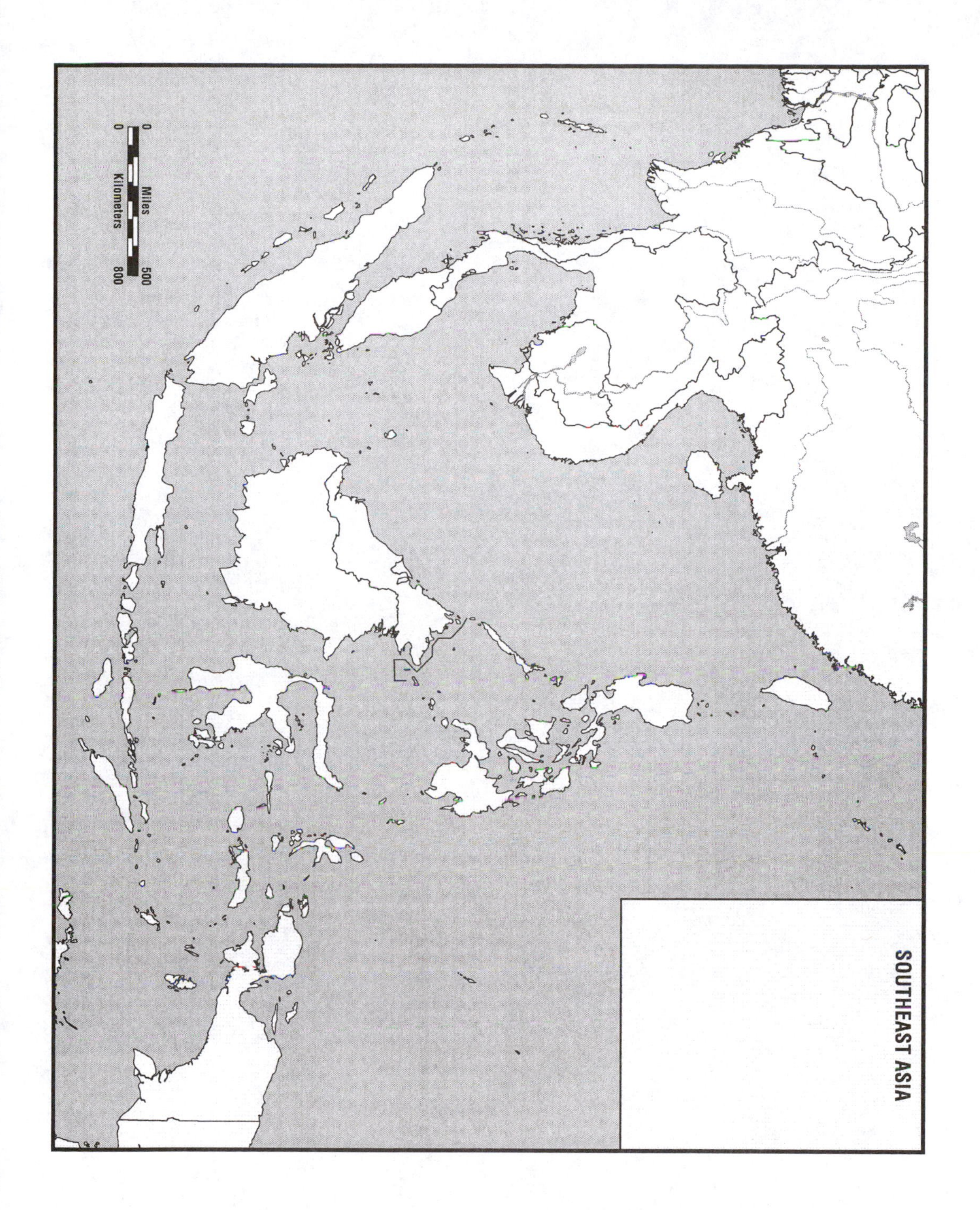
SOUTHEAST ASIA
0
Miles
500
0
Kilometers
800

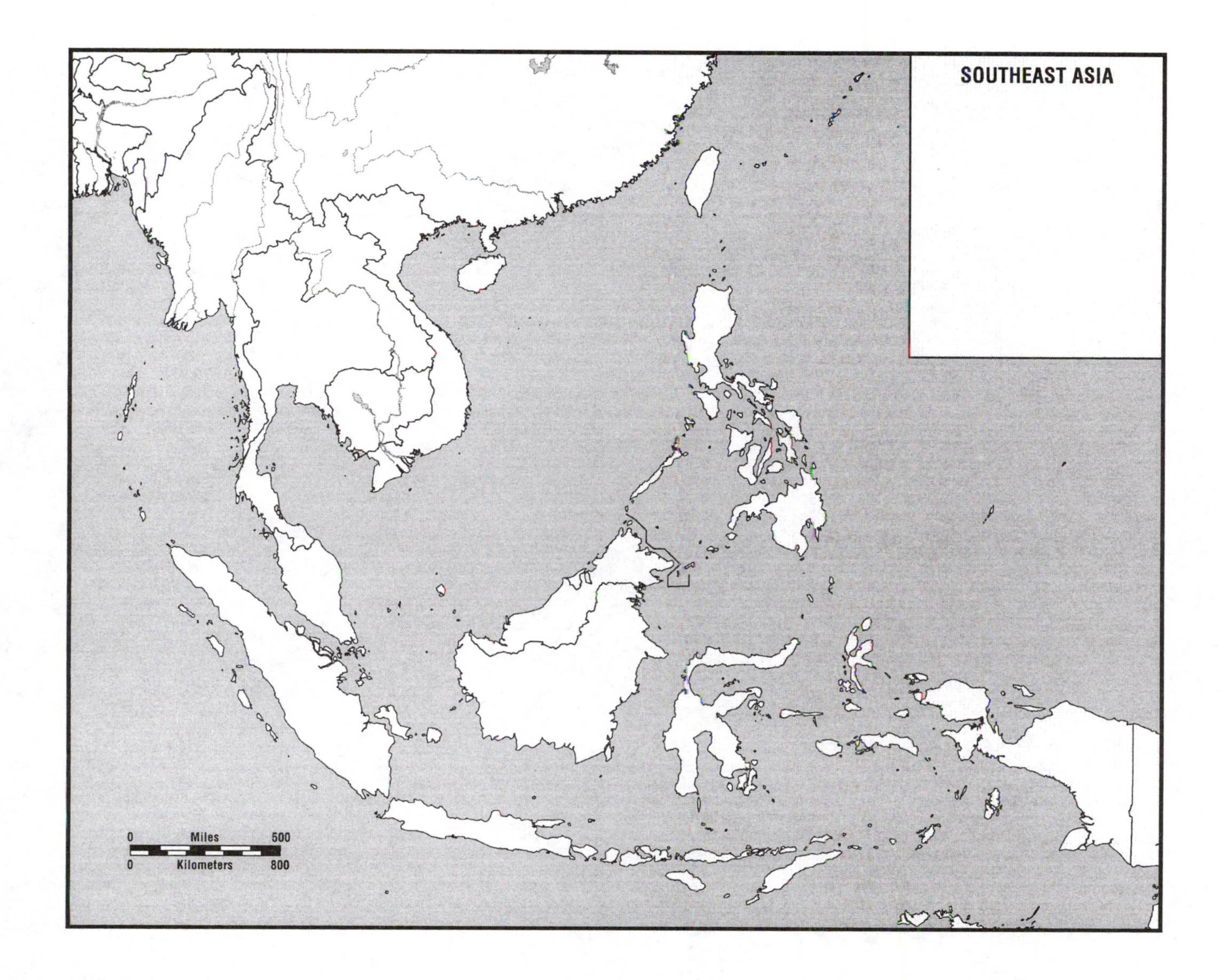
SOUTHEAST ASIA
0 Miles 500
0 Kilometers 800

CHAPTER ELEVEN

Oceania: Australia, New Zealand, and the Pacific

LEARNING OBJECTIVES

After reading the chapter and working through this study guide, you should:

- Know how island chains formed amid the vastness of the Pacific Ocean. Understand the role of the Pacific Ocean as a conduit for trade and cultural diffusion among island groups separated by vast distances of water.
- Know why the plant and animal life of Australia, New Zealand, and the far-flung islands are of such interest to scientists and social scientists. Know why recent accidentally introduced species have caused concern in the region.
- Know how the relationship of the region to the rest of the world, especially Asia and Europe, is shifting. Be able to explain the various reasons for this shift and possible impacts, both positive and negative.
- Know how the demographic and ethnic profiles of Australia, New Zealand, and the Pacific islands vary. Be aware of changing attitudes about ethnic diversity and understand the impacts of these changes on the social fabric.
- Understand the Pacific Way and how it serves as a unifying force to protect and promote the interests and culture of this relatively small population that inhabits such a large area.
- Relate current gender roles to those of the past. Understand how gender roles are changing in the region as a whole and within different ethnic groups.
- Know the state of environmental awareness and activism in the region and the environmental threats experienced. Know why the people of this region are relatively powerless to moderate these environmental threats. Know which threats have been imposed on Oceania by other world regions.
- Be able to explain why many Pacific islands can appear poor by standard measures (e.g., GDP per capita) yet provide their citizens with a reasonable level of affluence.

KEY TERMS

The following terms are in **bold** in the textbook. In the space next to the term (or on a separate sheet or flash cards), you can fill in the definitions for reference or quiz yourself for exam review. Definitions are found in the glossary of the textbook.

Aborigines

atoll

Austronesians

El Niño

endemic

Gondwana

Great Barrier Reef

high islands

hot spots

marsupials

Melanesia

Melanesians

Micronesia

MIRAB economy

monotremes

noble savage

ozone layer

Pacific Way

pidgin

Polynesia

roaring forties

subsistence affluence

REVIEW QUESTIONS

The following review questions are related directly to the textbook material. These questions can be used to help you prepare for an exam or you may want to read through the questions before you begin reading the textbook, quizzing yourself after you complete each section.

The Geographic Setting

1. Compare the geologic forces that account for the location and features of Australia, New Zealand, and the Pacific islands.
2. What are the differences between the geological forces that create low atolls vis-à-vis those that create high islands? How are *makatea* formed?
3. Compare the climate and moisture patterns of Australia and New Zealand.
4. What is the impact of the El Niño weather phenomenon on the islands in the Pacific?
5. Why are so many species of plants and animals endemic to certain places in Oceania?
6. Why are islands in Micronesia, Melanesia, and Polynesia classified into these three major groups? What role did historical migrations play in these classifications?
7. What types of Europeans first colonized Australia and New Zealand, and how did the origins of European settlers shift over time?
8. Define the term "Asianization". Describe how attitudes toward Asianization have changed in the last fifty years. Explain what has brought about such change.
9. Describe the two principal population patterns of this region and account for the differences.
10. Where is Australia's population concentrated? What accounts for this pattern?
11. List at least three key challenges faced by urbanizing areas and explain the root causes of these challenges.

Current Geographic Issues

12. What factors contribute to the new respect being shown for indigenous peoples of the region?
13. The Pacific Way seems to have started from a concern about using colonial-dominated school curricula. What does it encompass today?
14. Discuss several ways in which sports have been a unifying factor in the diverse and spacious region of Oceania.
15. New ways of life for women and men are emerging in this region. Discuss the varied historical role of gender in Oceania and how modern life is affecting what men and women do and the ways they interact.
16. List challenges and opportunities that tourism brings to this region. How are countries adapting to the challenges?
17. How did markets for Australia's and New Zealand's products change during the last half of the twentieth century? What accounts for the shift toward Asia? What were the economic impacts of these changes on Australia and New Zealand?
18. What is wrong with thinking that simply importing a natural predator of an exotic species can cure the ill effects of introducing that exotic species in the first place?

19. What human actions and global environmental crises have had a profound impact on the region's environmental status? How are the region's people responding?
20. What is subsistence affluence, and how does it insulate Pacific island economies from some of the advantages and disadvantages of global market economies?

CRITICAL THINKING EXERCISES

The following are "what if"-type questions that illustrate concepts you are learning from the textbook. These questions ask you to apply the ideas and principles you learned from the textbook to new situations.

1. Australia's immigration policies: Changes over time

Australia's immigrant population, if identified by century of migration to Australia, reflects distinct changes in the country's immigration policies.

- Draw a time line that is divided into the 1700s, 1800s, 1900s, and the beginning years of the twenty-first century.
- Indicate the years of the changes in Australia's immigration policies as specific points on your time line.
- Underneath the time line, indicate the top three source countries for immigration during each century.
- Evaluate the impact of immigration reform on Australia's population as a whole and identify impacts on various labor groups (e.g., professionals, blue-collar workers, and Aborigines) within Australian society.

2. Mobility in a far-flung region

If one were to characterize intraregional relationships and interactions for Oceania, the theme of mobility takes on some intriguing aspects.

- Select one of these areas: Polynesia, Micronesia, or Melanesia, and write an essay about the role of mobility in the history of that area.
- Discuss mobility in the twenty-first century for your selected area.
- Make some comparisons between your personal mobility patterns and the mobility patterns of the people in your selected area. What is different? What is the same?
- Finally, think in terms of impact. How does your personal mobility have an impact on your community or society? How does this compare with the impact that personal mobility in your selected area has on its community or society?

3. A new syncretic cultural tradition

The attitudes toward and roles of indigenous peoples are changing in this region. Indigenous ways are now blending with practices introduced by European colonization.

- Write an essay that argues for a syncretic (blending) cultural tradition that is currently evolving in this region.
- Provide explanations for what you see as the motivating factors for this change.
- Use and apply case studies from the Haka tradition and Aboriginal views of human-environment attachments.

- Reflect on your own community and identify possible cultural syncretism that you see taking place around you.

4. Technology in a region of great distances

Some suggest that technology may provide a valuable opportunity for this region to become more closely interconnected.

- Write an essay that identifies aspects of technology that could serve as vectors for communication and connection for this region. Check the CIA World Factbook for information on individual countries' technology foundations: www.cia.gov/cia/publications/factbook.
- In your essay, identify positive and negative impacts of the application of technology you chose for this region.
- Compare how your choices of technology for Oceania are similar or dissimilar to applications in your own community and life.

5. Living in the twenty-first century in a remote island nation

Place yourself as a citizen on one of the island nations discussed in this chapter. Imagine that you are a university student hoping to acquire adequate education to pursue a professional career.

- Based on your understanding of Oceania, suggest some of the assets that would be particular to you if you had grown up in a traditional island culture.
- Based on this same understanding of Oceania, suggest challenges that you would face in acquiring this education, searching for a related employment opportunity, and how such activities would affect your connection to your homeland.
- Suggest the adaptations that you would have to make as you complete your education and move into your professional career. Consider issues related to gender, a rural or semi-rural background, and indigenous cultural roots.

IMPORTANT PLACES

The following places are featured in the chapter. Make sure you can locate all of them on a map. Blank outline maps can be found on the textbook's Web site: www.whfreeman.com/pulsipher4e. Also, to prepare for quizzes and exams, write a few important facts about each place in the space provided.

Physical Features

1. Ayers Rock (Uluru)

2. Caroline Islands

3. Cook Islands

4. Coral Sea

5. Darling River

6. Easter Island

7. Eastern Highlands (Great Dividing Range)

8. Galápagos Islands

9. Great Australian Bight

10. Great Barrier Reef

11. Gulf of Carpentaria

12. Indian Ocean

13. Murray River

14. New Guinea

15. Pacific Ocean

16. Samoa Islands

17. Southern Alps

18. Tasman Sea

19. Tasmania

Regions/Countries/States/Provinces

20. American Samoa

21. Australia

22. Federated States of Micronesia

23. Fiji

24. French Polynesia

25. Guam

26. Hawaii

27. Kiribati

28. Marshall Islands

29. Melanesia

30. Micronesia

31. Nauru

32. New Caledonia

33. New Zealand

34. Northern Marianas

35. Palau

36. Papua New Guinea

37. Polynesia

38. Samoa (Western Samoa)

39. Solomon Islands

40. Tahiti

41. Tonga

42. Tuvalu

43. Vanuatu

Cities/Urban Areas

44. Adelaide

45. Apia

46. Auckland

47. Brisbane

48. Canberra

49. Christchurch

50. Funafuti

51. Honiara

52. Honolulu

53. Koror

54. Majuro

55. Melbourne

56. Newcastle

57. Nuku'alofa

58. Palikir

59. Papeete

60. Perth

61. Port Arthur

62. Port Moresby

63. Port Vila

64. Suva

65. Sydney

66. Tarawa

67. Wellington

68. Yaren

MAPPING EXERCISES

The following are three mapping exercises to improve your knowledge of the location of places, underscore why they are important, and clarify how they relate to each other. Some questions will ask you to locate places, compare maps, or fill in data; others will test your understanding of *why* you were asked to map the features that you did. Use the blank outline maps at the end of the chapter to complete these exercises. Additional blank outline maps can be found on the textbook's Web site: www.whfreeman.com/pulsipher4e.

1. Physical geography and population

The physical geography of Australia may play an important role in population distribution.

- Using Figure 11.1 (regional map), select five specific physical features (e.g., mountain ranges, coastal zones, and deserts) for Australia and delineate them on the blank map of Australia. Include an appropriate legend. Using Figure 11.8 (climate map), select three climatic zones and depict them on your map. Include an appropriate legend.
- Draw a dot and label cities with populations of 1 million or greater, as depicted in Figure 11.16 (population density map).

Questions

a. Justify your choice of physical features. Justify your choice of climatic zones.
b. Make at least two generalizations about Australia's population concentrations based on your selection of physical geography characteristics.
c. Make at least two generalizations about Australia's population concentrations based on your selection of climatic zones.
d. What are some unique challenges faced by people inhabiting regions with low population densities?
e. What are some challenges faced by people inhabiting regions with high population densities?

2. The United Nations Convention on the Law of the Sea and Oceania

The United Nations Convention on the Law of the Sea has particular impacts on this region.

- On a blank map of Oceania, approximate the 200-mile exclusive economic zone that each country can exploit. Draw this border around the island nations. You will have to make generalizations because of the scale at which you will be working.

Questions

a. The United Nations Convention on the Law of the Sea designates that countries sharing overlapping areas will draw a border midway between their coasts. Use a cross hatch marking (///) to identify any areas that may be in conflict. Suggest three possible conflicts that may arise.
b. Based on your understanding of Oceania, suggest benefits exclusive to these nations that are a result of the Law of the Sea Treaty.

c. Who do you think are the major global participants in exploiting the Exclusive Economic Zones? How can small independent island nations be assertive in defining cooperation with those who want to "lease" these zones?
d. Compare the gains and losses discussed above and decide the overall value of the Law of the Sea Treaty to Oceania.

3. **Global warming threatens Oceania**

A single meter rise in the sea level can have profound impacts on some of the islands and countries discussed in this chapter.

- Using a blank map of Oceania, color-code the islands according to their status of high island or low island. Refer to the CIA World Factbook at www.cia.gov/cia/publications/factbook for information about elevation minimums.

Questions

a. Suggest at least two impacts that both types of islands will experience if global warming results in a one meter rise in ocean levels. Identify at least two additional impacts that only the low islands will experience. In all cases, justify your reasons for selecting these particular impacts.
b. Identify strategies that these islands should follow to prevent catastrophic effects from global warming. What strategies are specific to high islands? What strategies are specific to low islands?

SAMPLE EXAM QUESTIONS

The following are sample questions to help you review for an exam. Answers are found in the back of this workbook.

1. Which of the following terms describes most of the animal life of present-day New Zealand?
   A. Endemic
   B. Extinct
   C. Native
   D. Introduced

2. The concept of the noble savage, created by romanticists and used by Pacific explorers, is most accurately described in which of the following statements?
   A. Pacific peoples are uncivilized and will wage war at any threat to their land.
   B. Pacific peoples are quite civilized on many of the same terms as Europeans.
   C. Pacific peoples live more civilized lives than Europeans do in dirty, crowded cities.
   D. Pacific peoples are primitive but untouched by corruption and moral debasement.

3. Which of the following describes the climate changes typically associated with an El Niño event in the western Pacific (Oceania)?
   A. Ocean temperatures *warm* up, which results in *more* cloud cover and rain.
   B. Ocean temperatures *cool* down, which results in *less* cloud cover and rain.
   C. Ocean temperatures *cool* down, which results in *more* cloud cover and rain.
   D. Ocean temperatures *warm* up, which results in *less* cloud cover and rain.

4. Which of the following represents one goal of the idea known in Oceania as the Pacific Way?
   A. To design a regional framework for Western-style development
   B. To be debt-free as a region by 2010
   C. To suppress industrial development during the colonial era
   D. To emphasize the unique regional identity of Oceania.

5. What type of language is often used by traders and others who live in the Pacific islands?
   A. Mandarin Chinese
   B. Pidgin
   C. Indo-European
   D. Islander

6. The interpretation of the terms of the Waitangi Treaty by the British went against which of the following Maori beliefs?
   A. Money was preferable to "guns and goods."
   B. Farming equipment would destroy the soil.
   C. Land would be returned at the turn of the century.
   D. Land is not a tradable commodity.

7. The modest size of the manufacturing sector in both Australia and New Zealand is the result of which one of the following?
   A. The unemployment rate is low, leaving few workers for the manufacturing sector.
   B. The UK has increased its reliance on the region's raw materials.
   C. The were unable to keep up with increasing domestic demand.
   D. Competition from Asian firms is stiff.

8. Which of the following statements provides the most accurate description of the composition of the populations of Australia and New Zealand since the 1990s?
   A. Ethnic diversity is increasing.
   B. The proportion of people of European descent is increasing.
   C. The population of people claiming indigenous roots is declining.
   D. The percentage of Asians is declining.

9. What feature lies off of Australia's northeastern coast and contributes to southeastern Australia's mild climate by directing warm west-flowing currents to the south?
   A. Great Australian Bight
   B. Easter Island
   C. Great Barrier Reef
   D. New Guinea

10. Which of the following best describes the population of the Pacific Islands in comparison to New Zealand and Australia?
    A. Older, less rapidly growing, and with life expectancies in the 80s
    B. Younger, more rapidly growing, and with life expectancies in the 60s or early 70s
    C. Younger, more rapidly growing, and with fertility rates trending rapidly higher
    D. Older, less rapidly growing, but with fertility rates trending rapidly higher

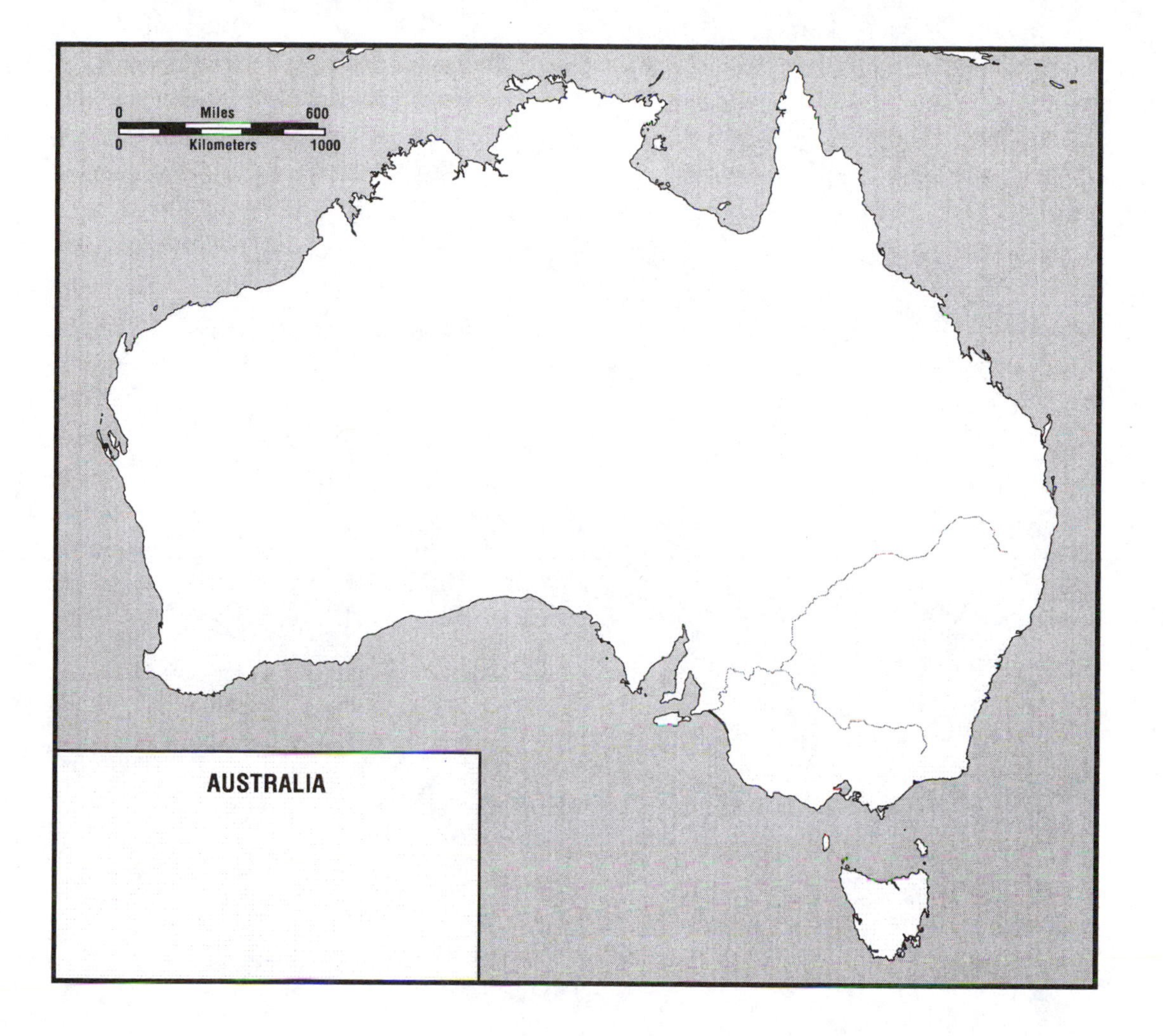
0 Miles 600
0 Kilometers 1000
AUSTRALIA

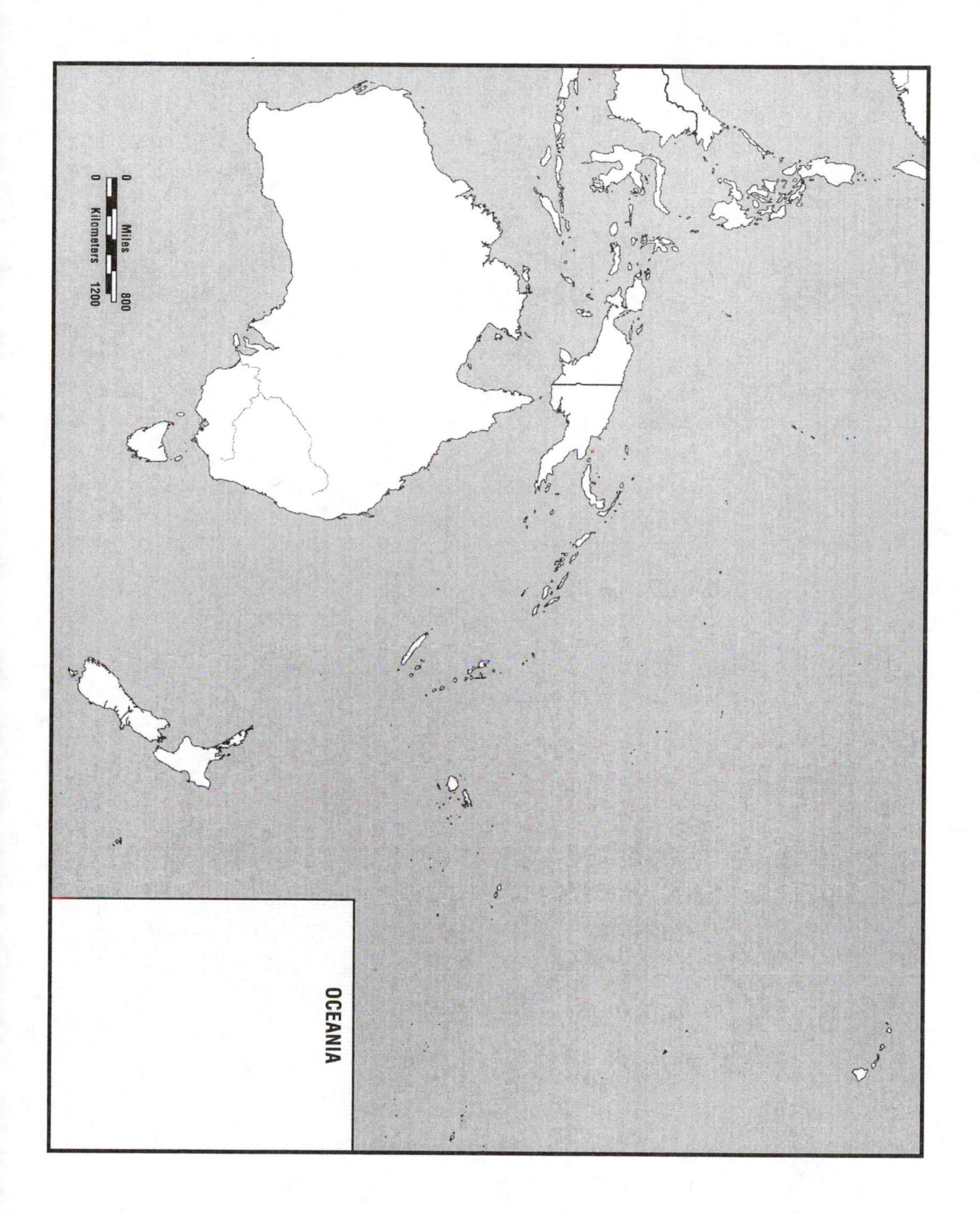
0 Miles 800
0 Kilometers 1200
OCEANIA

0 Miles 800
0 Kilometers 1200
OCEANIA

APPENDIX A

ANSWERS TO SAMPLE EXAM QUESTIONS

Chapter 1

1. C
2. D
3. C
4. B
5. D
6. D
7. D
8. A
9. A
10. B

Chapter 2

1. A
2. B
3. D
4. B
5. B
6. D
7. C
8. B
9. C
10. B

Chapter 3

1. B
2. A
3. C
4. C
5. B
6. C
7. B
8. D
9. D
10. A

Chapter 4

1. C
2. D
3. A
4. C
5. A
6. A
7. C
8. B
9. D
10. B

Chapter 5

1. B
2. B
3. C
4. D
5. C
6. B
7. B
8. B
9. D
10. A

Chapter 6

1. A
2. A
3. D
4. D
5. C
6. D
7. C
8. B
9. D
10. C

Chapter 7

1. D
2. B
3. D
4. A
5. D
6. A
7. A
8. A
9. D
10. D

Chapter 8

1. C
2. B
3. A
4. B
5. B
6. B
7. C
8. C
9. C
10. B

Chapter 9

1. C
2. A
3. D
4. C
5. C
6. A
7. B
8. D
9. C
10. C

Chapter 10

1. A
2. C
3. C
4. C
5. C
6. B
7. B
8. A
9. A
10. A

Chapter 11

1. D
2. D
3. B
4. D
5. B
6. D
7. D
8. A
9. C
10. B

APPENDIX B

BLANK WORLD MAPS

Additional blank maps can be found at: www.whfreeman.com/pulsipher4e

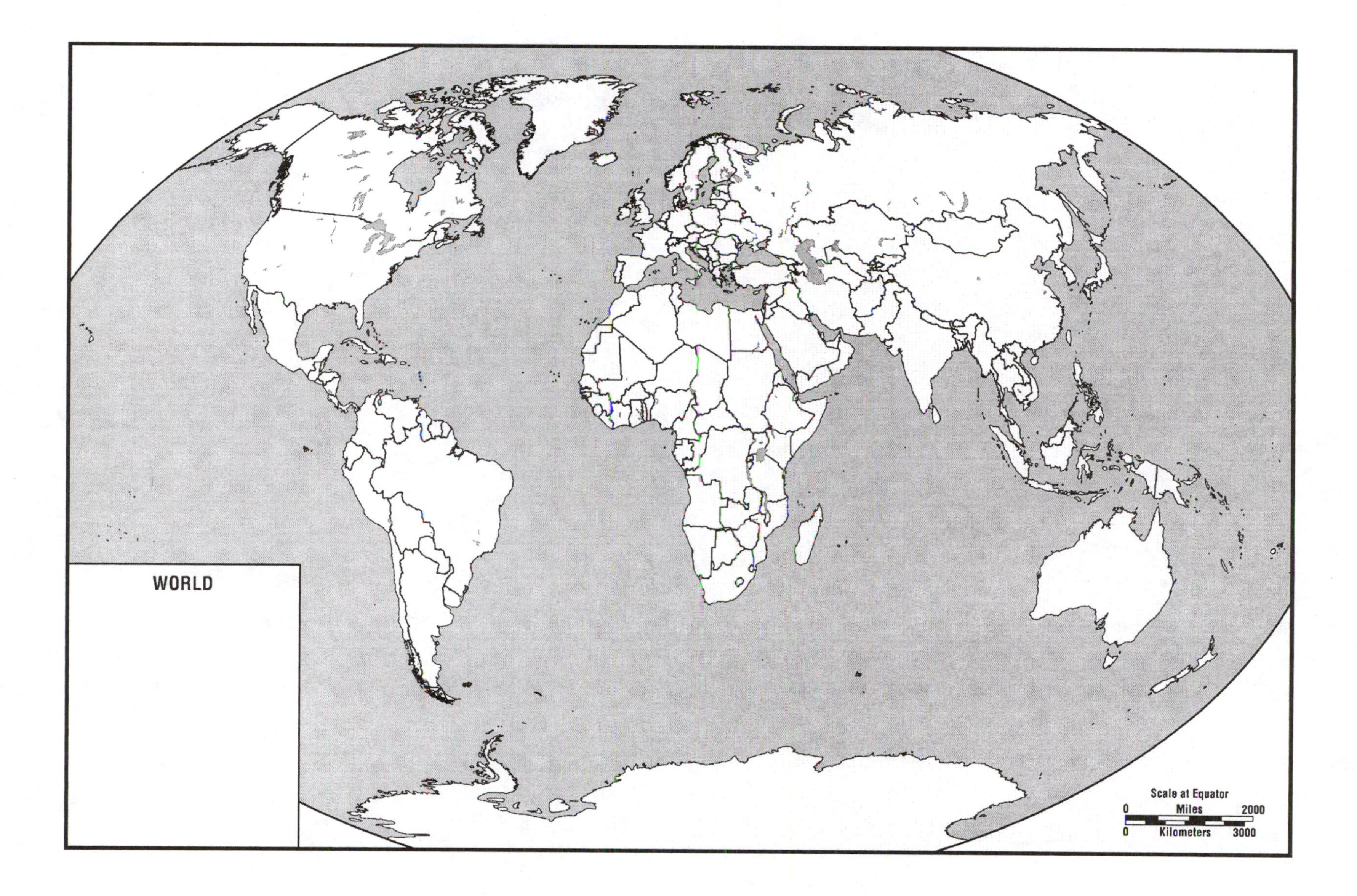
WORLD
Scale at Equator
0 Miles 2000
0 Kilometers 3000

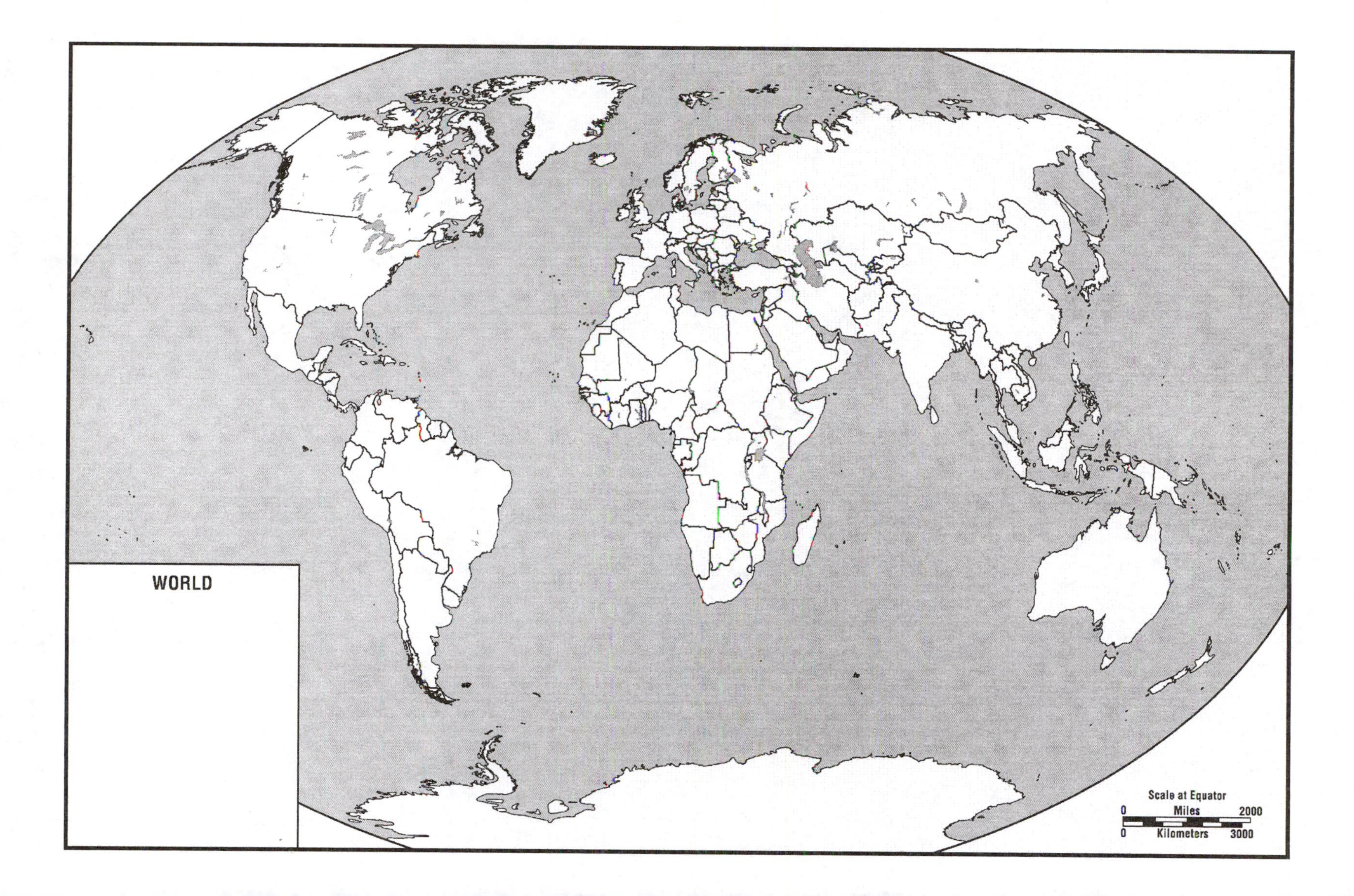
WORLD
Scale at Equator
0
Miles
2000
0
Kilometers
3000

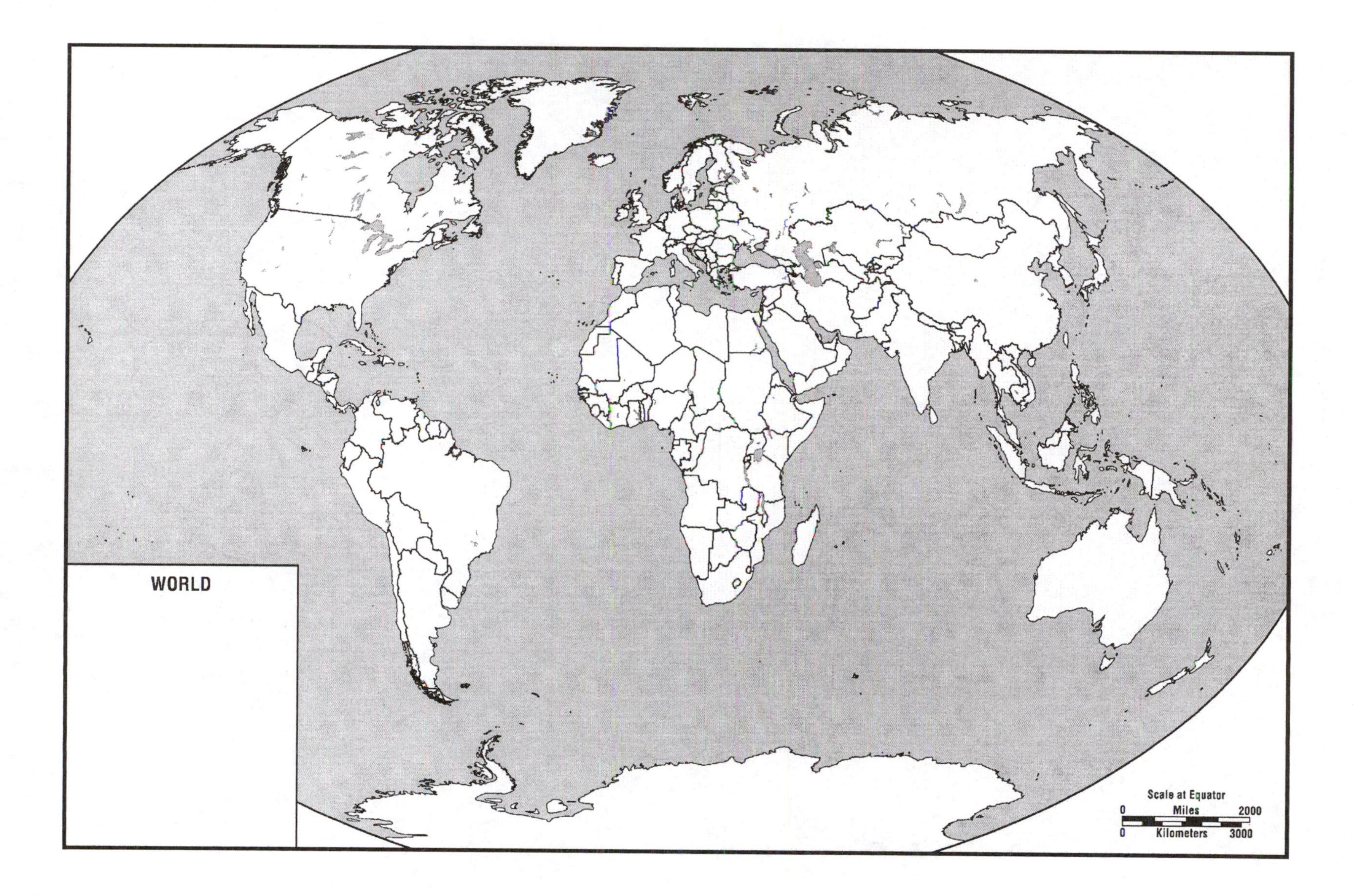
WORLD
Scale at Equator
0 Miles 2000
0 Kilometers 3000